Oxygen: High Enzymatic Reactivity of Reactive Oxygen Species

Authored by

Carmen Cecilia Espíndola Díaz

Oxygen: High Enzymatic Reactivity of Reactive Oxygen Species

Author: Carmen Cecilia Espíndola Díaz

ISBN (Online): 978-981-5036-63-3

ISBN (Print): 978-981-5036-64-0

ISBN (Paperback): 978-981-5036-65-7

©2021, Bentham Books imprint.

Published by Bentham Science Publishers Pte. Ltd. Singapore. All Rights Reserved.

need for a court order if at any point you breach any terms of this License Agreement. In no event will any delay or failure by Bentham Science Publishers in enforcing your compliance with this License Agreement constitute a waiver of any of its rights.

3. You acknowledge that you have read this License Agreement, and agree to be bound by its terms and conditions. To the extent that any other terms and conditions presented on any website of Bentham Science Publishers conflict with, or are inconsistent with, the terms and conditions set out in this License Agreement, you acknowledge that the terms and conditions set out in this License Agreement shall prevail.

Bentham Science Publishers Pte. Ltd.
80 Robinson Road #02-00
Singapore 068898
Singapore
Email: subscriptions@benthamscience.net

BENTHAM SCIENCE

CONTENTS

PREFACE i
    CONSENT FOR PUBLICATION ii
    CONFLICT OF INTEREST ii
    ACKNOWLEDGEMENTS ii

CHAPTER 1 OXYGEN 1
    INTRODUCTION 1
    1. NATURE 4
        1.2. OXYGEN FREE RADICALS 4
            1.2.1. Reactive Oxygen Species (ROS) 6
            1.2.2. Reactions of Reactive Oxygen Species in Aqueous Environments 8
            1.2.3. Fenton Reaction 14
        1.3. CYTOTOXICITY OF REACTIVE OXYGEN SPECIES (ROS) 16
            1.3.1. Superoxide 16
            1.3.2. Hydrogen Peroxide 18
            1.3.3. Hydroxyl Radical 18
            1.3.4. ROS Damage to Macromolecules 18
    CONCLUSION 23
    REFERENCES 23

CHAPTER 2 BIOLOGICAL OXIDATION 25
    INTRODUCTION 25
    1. REDOX POTENTIAL 25
        1.1. Oxidases 26
        1.2. NADH Dehydrogenase 29
        1.3. Hydroperoxidases 30
        1.4. Oxygenases 31
    2. MITOCHONDRIAL ELECTRON TRANSPORT CHAIN 33
    CONCLUSION 36
    REFERENCES 36

CHAPTER 3 REACTIVE OXYGEN SPECIES SOURCES 38
    1. ENDOGENOUS SOURCES 38
        1.1. Enzymes 38
            1.1.1. Cytochrome c – Enzymatic Oxidation 40
            1.1.2. Xanthine Oxidase (XO) -Reaction Center 46
            1.1.3. Xanthine Oxidoreductase (XOR) 54
            1.1.4. Galactose Oxidase 55
        1.2. CELLS AND ORGANELLES 65
            1.2.1. Phagocytic Cells 65
            1.2.2. Peroxysomes 67
            1.2.3. Mitochondria 67
            1.2.4. Microsomes 71
         1.3. Cytosolic Molecules 78
    2. EXOGENOUS SOURCES 78
        2.1. Haber – Weiss Reactions 79
        2.2. Iron 80
            2.2.1. Iron Storage Proteins 82
        2.3. COPPER 83
        2.4. CHROME 84
        2.5. CADMIUM 85

2.6. VANADIUM 85
2.7. MERCURY 86
2.8. NICKEL 86
2.9. ZINC 87
2.10. LEAD 87
2.11. COBALT 87
2.12. XENOBIOTICS. PARAQUAT – PQ 88
CONCLUSION 90
REFERENCES 91

CHAPTER 4 ANTIOXIDANT DEFENSE SYSTEMS 94
  INTRODUCTION 94
  1. PRIMARY ANTIOXIDANT DEFENSE SYSTEMS 94
    1.1. Enzymes 94
      1.1.1. Superoxide Dismutase-SOD 95
      1.1.2. Catalase 107
      1.1.3. Selenium-dependent Glutathione Peroxidase 108
      1.1.4. Glutathione Reductase 108
      1.1.5. Glucose-6-phosphate Dehydrogenase 108
      1.1.6. Other Enzymes 109
    1.2. Antioxidant Mitochondrial Defense 109
      1.2.1. Mitochondrial Coenzyme Q. Ubiquinone 113
    1.3. Non-enzymatic Traps 125
      1.3.1. Proteins 125
      1.3.2. Glutathione 126
      1.3.3. Vitamin C 128
  2. SECONDARY ANTIOXIDANT DEFENSE SYSTEMS 129
    2.1. ENZYMES 129
      2.1.1. Protein-specific Oxidoreductases 130
      2.1.2. Proteases 130
      2.1.3. Glutathione Peroxidase not Dependent on Selenium 130
      2.1.4. Phospholipases 131
    2.2. Non-enzymatic Traps 131
      2.2.1. Vitamin E 131
      2.2.2. Carotenoids 133
  3. ANTIOXIDANT PLANT DEFENSE SYSTEMS 134
  CONCLUSION 140
  REFERENCES 140

CHAPTER 5 FLAVONOIDS AS REACTIVE OXYGEN SPECIES PROMOTORS 144
  INTRODUCTION 144
  1. FLAVONOID CHEMICAL STRUCTURE 145
  2. TYPES AND SOURCES OF FLAVONOIDS 147
  3. FLAVONOIDS AS TRAPPING AGENTS FOR REACTIVE OXYGEN SPECIES -ROS. 149
    3.1. Structure-function Relationship 149
      3.1.1. Flavonoids as Antioxidants 149
    3.2. Redox Chemistry 155
  4. MECHANISMS OF FLAVONOID ANTIOXIDANT ACTIVITY 159
    4.1. Position and Number of Groups OH 161
    4.2. Formation Heat of Flavonoid Radicals (ΔHf) 164
    4.3. Bond Dissociation Energy (BDE) And Ionization Potential (IP) 165
      4.3.1. Quercetin 167

        4.3.2. Rutin 171
    4.4. Chelation of Metallic Ions 172
  5. INHIBITION OF PROOXIDANT ENZYMES MECHANISMS 173
    5.1. Xanthine Oxidase 173
    5.2. Lipoxygenase 175
  6. MEASUREMENT METHODS FOR FLAVONOID ANTIOXIDANT ACTIVITY 176
  7. SOME TECHNIQUES UTILIZED TO DETERMINE THE BIOAVAILABILITY OF
  FLAVONOIDS 180
  8. REACTIVE OXYGEN SPECIES MEASUREMENT 184
  9. INTESTINAL ABSORPTION AND PLASMA LEVELS OF FLAVONOIDS 186
  10. FLAVONOIDS PROOXIDANT ACTIVITY MECHANISM 193
  CONCLUSION 205
  REFERENCES 205

CHAPTER 6 OXYGEN AVAILABILITY 212
  1. REACTIVE OXYGEN SPECIES-ROS IN HYPOXIA 212
  CONCLUSION 217
  REFERENCES 217

SUBJECT INDEX 220

PREFACE

Due to the importance of oxygen to conserve and maintain the life of organisms on earth, it is imperative to be conscious of the need for knowledge about this element, its physical, chemical and physicochemical properties, metabolism, and everything related to its behavior and its relationship with living organisms in different ecosystems and environments. Similarly, it is vital to know the causes and serious consequences caused by the incorrect management of natural resources on the levels and quality of this element in the biosphere.

This book presents and analyses evidence of the high enzymatic reactivity of reactive oxygen species, their production sources, chemical formation mechanisms, enzymatic oxidation, reaction centers, mechanisms involved in oxidation-reduction reactions, cell respiration chemistry, enzymatic kinetics, electron transport chain mitochondrial and chloroplast, oxidation-reduction potential, reaction constants, reaction velocity and reaction mechanisms involved, cellular cytotoxicity, antioxidant defense mechanisms in plants and animals, the response of plants to conditions of environmental stress, xenobiotics, heavy metals, paraquat, and the thermodynamics inherent to oxygen metabolism. Chapter 5 presents evidence and analyzes the action of flavonoids as promoters of reactive oxygen species. It is written as a paradoxical example of the high reactive affinity of reactive oxygen species for enzymes since during the whole metabolic process that presents flavonoids as trapping agents of reactive oxygen species or oxidants, in the end, and due to this high affinity and reaction rates, they become promoting agents of the same reactive oxygen speciesi-ROS.

Dioxide O_2 is not stored in the body. However ambient air (or water) if it is the immediate reservoir of dioxide. The ability to extract oxygen from the environment and carry it to each cell in complex multicellular organisms through just-in-time metabolism was one of the main developments of organisms during evolution. In human cells, there is an increase in reactive oxygen species under conditions of low levels of available oxygen-hypoxia.

The unfortunate experience in which we human beings currently live has alerted all of humanity to the need to take care of nature and the need to have an environment that is as unpolluted as possible since there is sufficient scientific evidence to show the decrease in oxygen levels in the terrestrial and aquatic environments and the devastating effects this has on the survival of organisms. Therefore, there is a need to form citizen conscience about the care of nature and the presence of this essential element for life on earth.

CONSENT FOR PUBLICATION

Not applicable.

CONFLICT OF INTEREST

The author declares no conflict of interest, financial or otherwise.

ACKNOWLEDGEMENTS

Declared none.

Carmen Cecilia Espíndola Díaz

CHAPTER 1

Oxygen

Abstract: Earth's life depends mainly on the availability of oxygen in the terrestrial biosphere. Based on geochemical records of existing terrestrial oxides, oxygenic photosynthesis occurred in the cyanobacterial precursors approximately 2800 Ma ago. The oxygen level in the atmosphere is now 21%. The human cells use this oxygen to extract the necessary energy through mitochondrial respiration using the reactions of the redox system that involves the transfer of electrons, enzymatic agents, and reactive oxygen species, mainly superoxide radical (O^-_2), hydroxyl radical ($^\cdot OH$) and hydrogen peroxide (H_2O_2). The different reaction mechanisms from and to produce reactive oxygen species with their reaction constants in aquatic environments are presented here, as well as their production through the Fenton reaction. Oxidative stress is an imbalance between both normal oxygen-free radicals' production and the cell's ability to detoxify it.

Keywords: Enzymatic Reactivity, Great oxidation event-GOE, Hydrogen peroxide (H_2O_2), Hydroxyl radical ($^\cdot OH$), Lipid peroxidation, Oxygenic photosynthesis, Reactive oxygen species in aquatic environments, (ROS), ROS cytotoxicity, Superoxide anion (O^-_2).

INTRODUCTION

According to the geochemical records of terrestrial oxides that exist, the accumulation of oxygen in the Earth's atmosphere was originated from evolutionary processes that took place in the precursors of cyanobacteria before about 2.8 billion years ago [1].

There is evidence of a permanent increase in O_2 concentrations in the atmosphere since 2400 and 2100 Ma. Evidence of the presence of oxygen in the atmosphere is the appearance of soils with an oxidized red color and the disappearance of the old stream beds of easily oxidizable minerals such as pyrite (FeS_2) [2]. Oxygen constitutes 21% of the current atmosphere.

In the search for oxygen levels in the atmosphere, Lyon *et al.*, 2014, state that from the first photosynthetic production of oxygen and based on sulfur isotope records, after GOE, oxygen levels raised again and then decreased in the atmosphere where they remained for more than 1000Ma with very low levels. This extended inactivity was possibly caused by biogeochemical feedbacks that generated a deep ocean without oxygen. This anoxygenic ocean, with large deposits of H_2S attracted concentrations of bio-essential elements that, together with the low availability of oxygen, unchained the evolutionary events that gave rise to eukaryotic organisms and animals diversity until the final oxygenation and life expansion.

The differentiation of biotic or abiotic oxidation pathways, which can occur with or without oxygen, is the main difficulty in reaching a consensus on the appearance of atmospheric oxygen, despite intensive research in recent times.

Biomarkers are fossil molecules derived from organic compounds that bind to specific biological products present at the moment in which the sediments were deposited. The presence of cyanobacteria and eukaryotes in rocks from 2700 Ma ago was recorded with a biomarker [4].

The oldest producers of O_2 through photosynthesis, which are still found today, are cyanobacteria. Oxygen identification can also be performed by the recognition of sterane biomarkers in Eukaryotes since oxygen is required for the biological synthesis of sterols. This implies that the production and accumulation of oxygen occurred approximately 300Ma before the Great Oxidation Event - GOE, which occurred approximately 3800 to 2350 Ma ago.

The GEO is a time interval in which the differences in oxygen concentrations in the biosphere would represent a balance between early oxygen production and carbon deposits.

The available evidence suggests that oxygenic photosynthesis is much older than 2500Ma and that the production of oxygen through photosynthesis did not accumulate permanently in the atmosphere, due to the balance between carbon deposition and compensatory buffering [5].

There is evidence that, under conditions such as low SO_4 sulfate content in the archaic ocean and low O_2 levels in both the ocean and the atmosphere, high levels of methane (CH_4) and hydrocarbons from its photochemistry, such as ethane (C_2H_6), were produced.

O_2 levels in the earth are mainly due to photosynthesis. In the ocean, most of this oxygen is consumed through aerobic microbial respiration. In nature, the most complex metabolic process is oxygenic photosynthesis. It consists of two reaction centers in which electrons produced in the first reaction center (PSII) are then transferred to a second reducing center (PSI), through a cytochrome complex (Chapter 4).

A source of both light and electron reducing power is required for photosynthetic life. Considering that the electron donor for oxygenic photosynthesis is the surrounding water, it is possible that carbon fluxes through the biosphere were overloaded by oxygenic photosynthesis. This is supported because without an external source of carbon, H_2S-based photosynthesis is difficult to maintain, and organic matter deposits in archean reservoirs came from waters with Fe^{2+} levels.

It is unlikely that Fe^{2+} based photosynthesis would have occurred since this metabolism produces organic carbon particles and iron oxide minerals, which would disappear by microbial iron reduction. Likewise, H_2-based photosynthesis would also be unlikely to occur. Therefore, the most accurate explanation is that oxygenic photosynthesis was the origin of organic ponds in the pre-GOE ocean.

Oxygen present in nature has been originated by the fusion of ^4He atoms that occurs at high temperatures in stars, and the concentration of oxygen is approximately or higher than the concentration of carbon in the solar system. The electronic configuration of oxygen favors fast reactions with atoms and molecules to form radicals.

When oxygen reacts with a metal of groups I, II, III, IV, V, VI, corresponding oxides, such as H_2O, MgO, CaO, AlO, CO_2, SiO_2, NOx, PO_4, SOx are formed, and when it reacts with transition metals, such as Mn and Fe, it forms insoluble oxyhydroxides.

Redox reactions lead to oxygen reactivity and produce stable compounds such as H_2O, CO_2, HNO_3, H_2SO_4 and H_3PO_4 and intermediate unstable compounds, such as H_2O_2, NO, NO_2, CO, SO_2 are produced by abiotic oxygen reactions. Most oxygen reactions are exergonic.

Oxygen production from water oxidation is the most important reaction of oxygenic photosynthesis, in which Mn and Ca atoms are involved. By sequential electron transfer driven by a single photon, Mn atoms remove electrons releasing O_2 from the water molecule. Calcium stabilizes the intermediate oxygen until a second atom is released [6].

A complex five-step mechanism to remove four electrons and four protons (transition state S) is required to produce oxygen by water oxidation and is the most energy-demanding biological redox reaction [7].

1. NATURE

Oxygen is the most abundant element in the earth's crust and, after hydrogen and helium, is the third most common element in the universe. The discovery of oxygen has given rise to the understanding of similar chemical processes, such as combustion and aerobic catabolism: high-energy bonds that oxidize, releasing energy. In 1777 Lavoisier named this new element OXYGEN. His name means acid producer, so all acids were thought to contain this substance, and the understanding of oxygen chemistry by Lavoisier changed the concept of combustion theory by the concept of *oxidation*.

The most stable allotropic form of oxygen is dioxide O_2 and constitutes about 21% of the Earth's atmosphere being the key component of the two most important half-reactions for life on Earth:

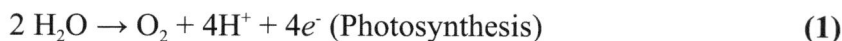

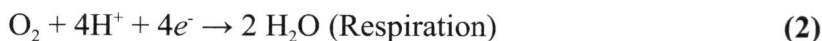

$$2\ H_2O \rightarrow O_2 + 4H^+ + 4e^- \text{ (Photosynthesis)} \tag{1}$$

$$O_2 + 4H^+ + 4e^- \rightarrow 2\ H_2O \text{ (Respiration)} \tag{2}$$

In reaction 1, energy from sunlight is captured to obtain protons and electrons that combine with CO_2 to produce (CHO)n, whose high-energy bonds are oxidized to form the carbon chemistry of photosynthesis for life. In reaction 2, the (CHO)n compounds are transformed to supply the energy needed for respiration. *These processes are carefully controlled by the enzyme systems of the cells.* As these protons and electrons are given to oxygen to form water, the energy of combustion is captured for synthesis, repair, and the work necessary for life.

For instance, in biological membranes, there are gaps such as the enzyme family's NADPH-oxidizes (Nox) that transfer electrons from NADPH, a two-electron reductant, to dioxygen to produce superoxide (reaction 3):

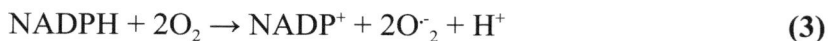

$$NADPH + 2O_2 \rightarrow NADP^+ + 2O^-_2 + H^+ \tag{3}$$

1.2. OXYGEN FREE RADICALS

A free radical is a chemical species that contains one or more missing electrons in its external orbital. Due to their electronic configuration, free radicals are unstable and extremely reactive since they quickly extract electrons from nearby molecules; so, they have a short half-life and a low steady-state concentration. The main types of biological reactions involving free radicals are presented in Table **1**.

Table 1. Main biological reactions of free radicals.

Reaction	Chemical Model
$A\cdot + B\cdot \rightarrow A\text{-}B$	Addition
$A\cdot + B\text{-}C\text{-}D\cdot \rightarrow A\text{-}B + C = D$	Disproportionation
$A\text{-}B\cdot \rightarrow A\cdot + B$	Fragmentation
$A\cdot + B\text{-}C \rightarrow A\text{-}B + C\cdot$	Transference of radical
$A\cdot + B = C \rightarrow A\text{-}B\text{-}C\cdot$	Addition

A non-free radical compound can become a free radical by the gain or loss of an electron. Free radicals can also easily form when a covalent bond is broken, leaving one electron of the shared pair in each of the atoms that were united; this process is called homolytic fission (reaction 4). Normally, when a bond is broken, it is broken heterolytically, *i.e.*, one of the atoms conserves both electrons, giving rise to an anion (negative charge), and the other atom loses an electron, giving rise to a cation (positive charge) (reaction 5).

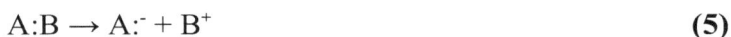

$$A{:}B \rightarrow A\cdot + B\cdot \tag{4}$$

$$A{:}B \rightarrow A{:}^- + B^+ \tag{5}$$

The oxygen molecule can be qualified as birradical since it has two missing electrons, each one located in a different anti-binding orbital $_\pi*$; this is the most stable state of oxygen and is called the ground state. Oxygen in its ground state, despite being powerful oxidants, is not very reactive. Due to the parallel spins of the unpaired electrons, the reactivity of oxygen as a biradical molecule decreases. Thus, when oxygen accepts a pair of electrons from another atom or non-radical molecule, these must-have parallel spins to the couple in *s* orbital vacancies $_\pi$5.

Considering the Pauli Exclusion Principle, the electron spins in an atomic or molecular orbital must have opposite directions; due to this, a restriction is imposed on an oxidation reaction by oxygen (Fig. **1**). Due to the spin constraint, oxygen reactions are slowed down, allowing electron transfer and free radical formation, this constitutes an advantage for aerobic organisms [8].

Molecular oxygen	↑	↑
Oxygen singlet delta ($^1\Delta gO_2$)	↑↓	
Oxygen singlet sigma ($^1\Sigma g^+O_2$)	↑	

Fig. (1). Arrangement of anti-bonding electrons $_\pi$* of oxygen [8].

An increase in the reactivity of molecular oxygen can be obtained to form singlet oxygen by spin inversion of one of the electrons of its outer electrons or by its sequential and univalent reduction to form oxygen free radical intermediates:

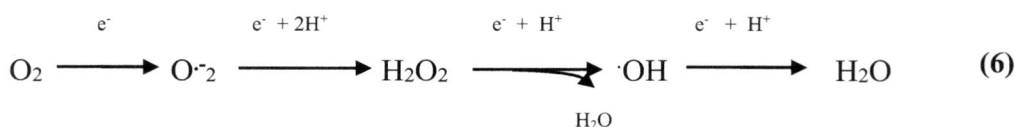

$$O_2 \xrightarrow{e^-} O^{\cdot-}_2 \xrightarrow{e^- + 2H^+} H_2O_2 \xrightarrow[H_2O]{e^- + H^+} {}^\cdot OH \xrightarrow{e^- + H^+} H_2O \qquad (6)$$

1.2.1. Reactive Oxygen Species (ROS)

The following table lists the main reactive oxygen species produced in biological systems.

Table 2. Main reactive oxygen species.

ROS	Name
$O^{\cdot-}_2$	Superoxide anion radical
$HO^{\cdot}_2$	Perhydroxyl radical
H_2O_2	Hydrogen peroxide
$^\cdot OH$	Hydroxyl radical
$RO^{\cdot}$	Alkoxy radical
$ROO^{\cdot}$	Peroxyl radical
$^1\Delta gO_2$	Oxygen singlet delta
O_2	Molecular oxygen

1.2.1.1. Singlet Oxygen

The parallel spins of the two electrons of the external orbitals of molecular oxygen can become anti-parallel by a pulse of energy, resulting in singlet oxygen. There are two types of singlet oxygen: the oxygen singlet delta ($^1\Delta gO_2$), which is the most biologically important due to its long half-life and oxygen singlet sigma ($^1\Sigma^+O_2$), very reactive, but with a short half-life because, after forming, it quickly decays to the singlet delta oxygen state. The molecular oxygen to singlet oxygen excitation can be carried out by several biological pigments such as chlorophyll or retinal when illuminated with light of a certain wavelength in O_2 presence. The pigments absorb the light, enter a higher state of electronic excitation, and then transfers energy to O_2 to form singlet oxygen while returning to its original state.

1.2.1.2. Superoxide Radical

The superoxide radical ion ($O^{\cdot-}_2$) is formed when the molecular oxygen is reduced by an electron. This chemical species is very reactive and very unstable in aqueous solutions since can react spontaneously to herself by a dismutation reaction to produce hydrogen peroxide (H_2O_2) and molecular oxygen (reaction 7). With neutral or physiological pH, this dismutation reaction is catalyzed by superoxide dismutase (SOD). Singlet oxygen can also be formed during superoxide dismutation, however, only less than 0,008% of the oxygen produced in this way is in the singlet state.

$$O^{\cdot-}_{2+}\, O^{\cdot-}_2 + 2H^+ \rightarrow H_2O_2 + O_2 \tag{7}$$

The superoxide radical, at low pH, can be in its proton form as a perhydroxyl radical ($HO^{\cdot}_2$), from which hydrogen peroxide is rapidly formed (reaction 8).

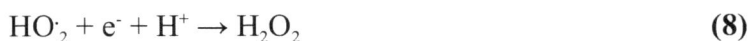

$$HO^{\cdot}_2 + e^- + H^+ \rightarrow H_2O_2 \tag{8}$$

1.2.1.3. Hydrogen Peroxide

When two electrons reduce the oxygen molecule, peroxide ion is produced (O_2^{2-}), whose protonated form is hydrogen peroxide. H_2O_2 is not a free radical and generally, in aqueous media, it does not oxidize some organic molecules; however, it is a biologically important oxidant because from it, through its interaction with transition metals, the hydroxyl radical ($^{\cdot}OH$), is generated. The H_2O_2 is dangerous for cells because it is generally not ionized and can spread through cell membranes.

1.2.1.4. Hydroxyl Radical

The reduction of molecular oxygen by three electrons originates the free radical hydroxyl (see also Fig. 2, chapter 4). This is a highly reactive chemical species that can react with any biological molecule at a rate of 10^7- 10^{10} mol/seg; therefore, its half-life and radius of action are extremely short (fractions of microseconds and 30 Å, respectively). The main source of hydroxyl radicals is the *Haber-Weiss* reaction (reaction 11), which results from the balance of two reactions (reactions 9 and 10), the second of which is Fenton's reactions which requires an iron chelate to produce. Other transition metals, similarly, accelerate the production of hydroxyl radicals.

$$O^{-}_{2} + Fe^{3+} \rightarrow O_2 + Fe^{2+} \tag{9}$$

$$Fe^{2+} + H_2O_2 \rightarrow Fe^{3+} + OH^- + \cdot OH \tag{10}$$

$$O^{-}_{2} + H_2O_2 \rightarrow O_2 + OH^- + \cdot OH \tag{11}$$

1.2.2. Reactions of Reactive Oxygen Species in Aqueous Environments

Ionizing radiations, ultraviolet radiations and particular radiations transfer their energy to cell components and are therefore a source of free radicals. These radiations produce the heterolytic fission of water to obtain hydrated electrons, hydrogen atoms, superoxide, hydroxyl radicals, and hydrogen peroxide, in the presence of oxygen. Free radicals can also be generated by the photolysis of chemical bonds caused by visible light of a suitable wavelength, mainly in the presence of photosensitizers.

In water radiolysis, when a short pulse of electrons is applied on pure water after 10^7 s, the processes are more direct and the reactive intermediates: hydrated electron, hydrogen atom, hydroxyl radical, hydrogen peroxide and molecular oxygen are homogeneously distributed in the solution. Their products respectively are: $G(e^-_{aq}) = 0.27$ µmolJ^{-1}, $G(H^\cdot) = 0.07$ µmolJ^{-1}, $G(\cdot OH) = 0.27$ µmolJ^{-1}, $G(H_2O_2) = 0.07$ µmolJ^{-1}. When studying the reaction of a given intermediate with the solute S, concentrations of solute high enough to ensure the complete activity of the investigated intermediate are utilized. This concentration is determined by the products of the reaction constants (k) and the concentration of the solute [S]. This quantity is called the trapping capacity = k[S].

When the proper concentration is selected, the concentrations of the intermediate states formed in both the solute and the water radiolysis are practically equal. For example, for the reaction of the hydrated electron, we have $G(e^-_{aq}) = G(S^-)$. The value of G is utilized to calculate the molar absorption coefficient, which is necessary to calculate the second-order constant and for the identification of the absorption spectra.

With a constant k in the controlled diffusion range $\sim10^{10}mol^{-1}dm^3s^{-1}$ and with a concentration of solute 10^{-3} moldm^{-3} [S], the entrapment is nearing completion (trapping capacity, $k[S] \approx 10^7 \ s^{-1}$). However, in practice, the trapping capacity is lower than 10^7s^{-1} e.g. due to low reactivity (low k) or low solubility (low [S]). In the case of molecules with a strong absorbance in the near UV/visible region (low light transparency), low solute concentrations could be utilized.

For example, it is possible to consider organic dyes with molar absorption coefficients greater than $10^5mol^{-1}dm^3 \ cm^{-1}$ in the visible range. The spectra of these molecules overlap strongly with those of the intermediate states produced. The net result can be negative or positive. Because a decrease in absorbance due to depletion (Δ negative absorbance) or an increase in absorbance due to intermediate states (Δ positive absorbance) could occur.

When the trapping capacity is less than 10^7s^{-1}, the intermediate + solute reaction of water radiolysis is not complete and an amount of solute is lost in each concentration of the water radiolysis intermediates or in their reactions with H_2O, H^+ and H_2O_2.

Pálfi *et al.*, 2010, made kinetic calculations of the reactions that occur in water radiolysis, based on the following 50 reactions in Table **3**.

Table 3. Reactions of Oxygen free radicals and their constants (mol^{-1}dm^{3}s^{-1}).

Reaction No	Reaction	Constants
12	$e^-_{aq} + e^-_{aq} + 2H_2O \rightarrow H_2 + 2OH^-$	5.5×10^9
13	$e^-_{aq} + H^{\cdot} + H_2O \rightarrow H_2 + OH^-$	2.31×10^{10}
14	$e^-_{aq} + {\cdot}OH \rightarrow OH^-$	3×10^{10}
15	$e^-_{aq} + O^{\cdot-} + H_2O \rightarrow 2OH^-$	2.2×10^{10}
16	$e^-_{aq} + H_2O_2 \rightarrow {\cdot}OH + OH^-$	1.1×10^{10}
17	$e^-_{aq} + HO_2^- \rightarrow O^{\cdot-} + OH^-$	3.5×10^9
18	$e^-_{aq} + O_2 \rightarrow O^{\cdot-}_2$	1.9×10^{10}
19	$e^-_{aq} + O^{\cdot-}_2 \rightarrow O_2^{2-}$	1.3×10^{10}
20	$e^-_{aq} + HO_2^{\cdot} \rightarrow HO_2^-$	1.7×10^{10}
21	$e^-_{aq} + H_2O \rightarrow H^{\cdot} + OH^-$	1.9×10^{10}
22	$e^-_{aq} + H_3O^+ \rightarrow H^{\cdot} + H_2O$	2.3×10^{10}
23	$H^{\cdot} + H^{\cdot} \rightarrow H_2$	7.8×10^9

(Table 3) cont.....

Reaction No	Reaction	Constants
24	$H^. + {^.}OH \rightarrow H_2O$	7×10^9
25	$H^. + H_2O_2 \rightarrow {^.}OH + H_2O$	5.1×10^7
26	$H^. + HO_2{^-} \rightarrow {^.}OH + HO^-$	1.2×10^9
27	$H^. + O_2 \rightarrow HO_2{^.}$	2.1×10^{10}
28	$H^. + HO_2{^.} \rightarrow H_2O_2$	1.92×10^{10}
29	$H^. + O^.{_2} \rightarrow HO_2{^-}$	1.92×10^{10}
30	$H^. + OH^- \rightarrow e^-{_{aq}} + H_2O$	2.5×10^7
31	${^.}OH + {^.}OH \rightarrow H_2O_2$	5.5×10^9
32	${^.}OH + O^.{^-} \rightarrow HO_2{^-}$	1.73×10^{10}
33	${^.}OH + H_2O_2 \rightarrow H_2O + HO_2{^.}$	2.7×10^7
34	${^.}OH + HO_2{^-} \rightarrow H_2O + HO_2{^.}$	7.5×10^9
35	${^.}OH + HO_2{^.} \rightarrow H_2O + O_2$	6.8×10^9
36	${^.}OH + O^.{_2} \rightarrow OH^- + O_2$	9.52×10^9
37	${^.}OH + H_2 \rightarrow H^. + H_2O$	4.2×10^7
38	$O^.{^-} + O^.{^-} \rightarrow O_2{^{2-}}$	2×10^9
39	$O^.{^-} + H_2O_2 \rightarrow H_2O + O^.{_2}$	5×10^8
40	$O^.{^-} + HO_2{^-} \rightarrow OH^- + O^.{_2}$	4×10^8
41	$O^.{^-} + O_2 \rightarrow O_3{^.}{^-}$	3.6×10^9
42	$O^.{^-} + O^.{_2} + H_2O \rightarrow 2OH^- + O_2$	6×10^8
43	$O^.{^-} + H_2 \rightarrow H^. + HO^-$	1.1×10^8
44	$H_2O_2 \rightarrow H_2O + O^.$	$3.36 \times 10^{-8\,a}$
45	$O^. + O^. \rightarrow O_2$	4.8×10^9
46	$HO_2{^.} + HO_2{^.} \rightarrow H_2O_2 + O_2$	8.3×10^5
47	$HO_2{^.} + O^.{_2} \rightarrow HO_2{^-} + O_2$	9.7×10^7
48	$O^.{_2} + O^.{_2} \rightarrow O_2{^{2-}} + O_2$	1.42×10^9
49	$O_2{^{2-}} + H_2O \rightarrow HO_2{^-} + OH^-$	9.61×10^5
50	$OH^- + H_3O^+ \rightarrow 2H_2O$	7.5×10^{10}
51	$2H_2O \rightarrow OH^- + H_3O^+$	2.5×10^{-7}
52	${^.}OH + OH^- \rightarrow O^.{^-} + H_2O$	1.2×10^{10}
53	$O^.{^-} + H_2O \rightarrow {^.}OH + OH^-$	9.3×10^7
54	$H_2O_2 + OH^- \rightarrow HO_2{^-} + H_2O$	1.3×10^{10}
55	$HO_2{^-} + H_2O \rightarrow H_2O_2 + OH^-$	9.8×10^5
56	$O^.{_2} + H_3O^+ \rightarrow HO_2{^.} + H_2O$	4.8×10^{10}
57	$HO_2{^.} + H_2O \rightarrow O^.{_2} + H_3O^+$	1.3×10^4

(Table 3) cont.....

Reaction No	Reaction	Constants
58	$HO_3{}^{\cdot} \rightarrow {}^{\cdot}OH + O_2$	1.1×10^{5a}
59	$O_3{}^{-} \rightarrow O^{\cdot-} + O_2$	1.8×10^{6a}
60	$O_3{}^{\cdot-} + H_3O^+ \rightarrow HO_3{}^{\cdot} + H_2O$	5.2×10^{10}
61	$HO_3{}^{\cdot} + H_2O \rightarrow O_3{}^{\cdot-} + H_3O^+$	6.7×10^{2}
62	$e^-{}_{aq} + N_2O \rightarrow O^{\cdot-} + N_2$	9.1×10^{9}
63	$H^{\cdot} + N_2O \rightarrow {}^{\cdot}OH + N_2$	2.1×10^{6}
64	${}^{\cdot}OH + (CH_3)_3COH \rightarrow H_2O + {}^{\cdot}CH_2(CH_3)_2COH$	4×10^{9}
65	$H^{\cdot} + (CH_3)_3COH \rightarrow H_2 + {}^{\cdot}CH_2C(CH_3)_2OH$	1×10^{6}

[a] s^{-1}

From the studies performed, the concentration of $e^-{}_{aq}$, is maintained in the range of 5-10 at a constant dose/pulse value. However, the concentration of $e^-{}_{aq}$ decreases at a pH at which the ${}^{\cdot}H$ atom increases due to $H/e^-{}_{aq}$ conversion in the reaction (reaction 22), for example, at pH < 5.

In the reaction $H^{\cdot} + OH^-$ (reaction 30) $e^-{}_{aq}$ production increases at pH > 13, has a relatively low coefficient of 2.5×10^7 $mol^{-1}dm^3s^{-1}$ and is an important reaction at high OH^- concentrations. The ${}^{\cdot}OH$ concentration reflects the conversion ${}^{\cdot}OH/O^{\cdot-}$ (reaction 52) with pK_a of 11.9 and H_2O_2 concentration reflects the conversion $H_2O_2/HO_2{}^-$ (reaction 55) with pK_a of 11.6 [9].

Considering the sensitivity analysis in pure water performed, of the 50 reactions, only 21-31 reactions considerably influenced the concentrations of intermediates in neutral solutions. The hydrated electrons concentration depended on largely on self-reaction (reaction 12). In such a manner, that in the reactions 13, 14 and 16, the $e^-{}_{aq}$ reaction with $H^{\cdot}$, ${}^{\cdot}OH$ and H_2O_2 contribute very few to the deterioration. The concentration of the hydrogen atom depends from the reaction of the ${}^{\cdot}H$ atom with the hydrated electron (reaction 13), of the reaction with another hydrogen atom (reaction 23) and from the reaction with the hydroxyl radical (reaction 24). For hydroxyl radical, the self-reaction (reaction 31) and the reaction with the hydrated electron (reaction 14) are of particular importance.

For the hydrated electron reactions, calculations were made on solutions with 5% v/v tert-butanol. Selected doses/pulse of 1, 10 and 100 Gy, was also applied 1×10^9 $mol^{-1}s^{-1}$ for the coefficients of the solutes with the hydrogen atom (reaction 67).

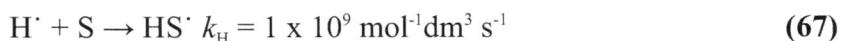

$$e^-{}_{aq} + S \rightarrow S^- \tag{66}$$

$$H^{\cdot} + S \rightarrow HS^{\cdot} \quad k_H = 1 \times 10^9 \ mol^{-1}dm^3 \ s^{-1} \tag{67}$$

With one dose/pulse of 1-100 Gy the hydrated electrons concentration at 10^{-7}s is in the range of 10^{-5} - 10^{-7} moldm^{-3}.

At high trapping capacities $k[S] = 10^7$s^{-1} (example, with $k \approx 1$ x 10^{10} mol^{-1}dm^{3}s^{-1} and $[S] \approx 10^{-3}$ moldm^{-3}), react with the solute all the electrons released. With doses 10 Gy/pulse at a solute concentration of 10^{-6} moldm^{-3} only 10% of the e^-_{aq} disappear in the reaction with S. The controlled diffusion-reaction 66 is similar to that of reactions 12, 13 and 14.

When $k = 1$ x 10^{10} mol^{-1}dm^{3}s^{-1} to obtain SF = 0.5, $[S] = 3.4$ x 10^{-6}, 3.4 x 10^{-5} and 3.4 x 10^{-4} moldm^{-3}, is necessary to 1, 10 and 100 Gy doses/pulse, respectively. The formation rate of this depends directly on the solute concentration.

Pseudo-first order conditions were considered to determine the reaction coefficient e^-_{aq} + S, in pulse radiolysis measurements. The calculations are based on the constant $k_H = 1$ x 10^9 mol^{-1}dm^{3}s^{-1} for reactions of the hydrogen atom with S (reaction 67). When changed k_H Pálfi *et al.*, 2010, found that its value slightly influences on S$^-$ production and that in contrast, the production of HS$^{\cdot}$ in reaction 67 is strongly influenced by ke^-_{aq}.

$$SF = [1 + 2 \text{ x } e^{\lambda(\chi - \chi_c)}]^{-1/2} \qquad \textit{(Equation 1)}$$

$$\chi = \log (k[S]/D) \qquad \textit{(Equation 2)}$$

Where λ and χ_c are constants (Table **4**) [9]. According to this *equation*, when $k[S]$ is very low, slightly trapping product is formed and when the value of $k[S]$ is high, the SF is near to unity.

Table 4. Parameters obtained by adjustment.

	λ (log(s Gy))	χ_c (log (s^{-1} Gy^{-1}))
e^-_{aq}	4.265 ± 0.015	4.617 ± 0.001
$^{\cdot}$OH	4.358 ± 0.011	4.665 ± 0.001

Under oxidative conditions (saturated solution of N$_2$O, $[N_2O] = 2.5$ x 10^{-2} moldm^{-3} at 1 pressure atmosphere) for hydroxyl radical reactions, similar trends in the formation of $^{\cdot}$SOH were observed, as when observed under reducing conditions to produce S$^-$ in e^-_{aq} trapping.

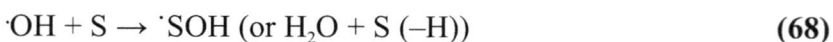

$$^{\cdot}OH + S \rightarrow {}^{\cdot}SOH \text{ (or } H_2O + S \text{ (–H))} \qquad \textbf{(68)}$$

The reaction (31) $\cdot OH + \cdot OH$ which has the same constant 5.5×10^9 mol^{-1}dm^{3}s^{-1} that the reaction (12) $e^-_{aq} + e^-_{aq}$ is the main competitor of the reaction (68) $\cdot OH + S$.

To compare the value of the equation with the parameters obtained by adjustment, pulse radiolysis experiments were performed at room temperature. In the reactions of $\cdot OH$ with thiocyanate ion and with *p*-cresol, the absorbance of the $\cdot OH$ radical was measured, depending on the concentration of this. With electron trapping utilizing diethyl fumarate, e^-_{aq} reaction was studied. Applying the dose values of selected constants, the concentration of the solute varied between 10^{-6} and 10^{-3} moldm^{-3} and the absorbance of the product was determined.

The determination of the relative product yield is facilitated by an electron adduct with a very high molar absorption coefficient 17,000 mol^{-1}dm^{3}cm^{-1} at 335 nm, which occurs when hydrated electrons react with diethyl fumarate.

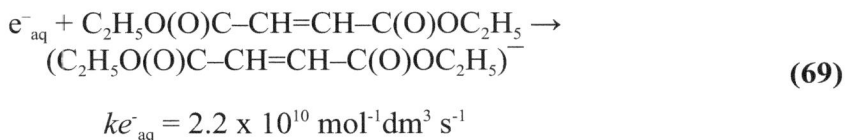

$$e^-_{aq} + C_2H_5O(O)C-CH=CH-C(O)OC_2H_5 \rightarrow$$
$$(C_2H_5O(O)C-CH=CH-C(O)OC_2H_5)^- \quad \quad \text{(69)}$$

$$ke^-_{aq} = 2.2 \times 10^{10} \text{ mol}^{-1}\text{dm}^3 \text{ s}^{-1}$$

In the $\cdot OH$ reaction with SCN$^-$ ion, well-measured temporary products (SCN)$_2^{\cdot-}$ are formed in two-step processes, the second step involves balancing:

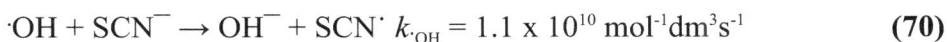

$$\cdot OH + SCN^- \rightarrow OH^- + SCN^\cdot \quad k_{\cdot OH} = 1.1 \times 10^{10} \text{ mol}^{-1}\text{dm}^3\text{s}^{-1} \quad \text{(70)}$$

$$SCN^\cdot + SCN^- \leftrightarrow (SCN)_2^{\cdot-} \quad k_{SCN^\cdot} = 2.0 \times 10^{10} \text{ mol}^{-1}\text{dm}^3\text{s}^{-1} \quad \text{(71)}$$

$K = 2 \times 10^5$ moldm^{-3}

The measurement of *p*-cresol was performed at neutral pH, observing the increase in absorbance around 435 nm of the cyclohexadienyl type radical [9]. Furthermore, its transformation into phenoxyl radical was slow and the reaction occurred in about 100 µs.

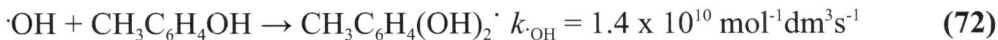

$$\cdot OH + CH_3C_6H_4OH \rightarrow CH_3C_6H_4(OH)_2^\cdot \quad k_{\cdot OH} = 1.4 \times 10^{10} \text{ mol}^{-1}\text{dm}^3\text{s}^{-1} \quad \text{(72)}$$

$$CH_3C_6H_4(OH)_2^\cdot \rightarrow CH_3C_6H_4O^\cdot + H_2O \rightarrow k_{elim} = 5 \times 10^4 \text{ s}^{-1} \quad \text{(73)}$$

(pH 5-7)

By developing the empirical *equation*, it is possible to estimate the production of intermediate products, which are utilized for the absorption spectra correction and for the extinction coefficients calculation.

1.2.3. Fenton Reaction

The classical Fenton mechanism (reaction 74, or with ligands reaction 75) determines that free hydroxyl radicals are produced when hydrogen peroxide is reduced at an iron center.

$$Fe^{2+} + H_2O_2 \rightarrow Fe^{3+} + OH^- + \cdot OH \qquad (74)$$

$$Fe^{II}(H_2O)_6^{2+} + H_2O_2 \rightarrow [Fe^{III}(H_2O)_5OH]^{2+} + \cdot OH + H_2O \qquad (75)$$

Some hydroxyl radicals produced in the Fenton reaction, however, remain attached to the iron center as $[Fe....OH]^{3+}$ or $[Fe{=}O]^{2+}$ intermediates (reaction 76).

$$Fe^{2+} + H_2O_2 \rightarrow [Fe{=}O]^{2+} + OH^- \qquad (76)$$

When an analogy is established between the reaction kinetics of hydroxyl radicals originating from the fenton reaction and those originating independently of iron, it is found that the intermediates from reaction 76 have different oxidizing properties.

Research performed on Fenton reagents, it was concluded that ·OH is not the dominant reagent and that a nucleophilic adduct reacts directly with the substrates, with the presence of an iron chelator.

For reactions involving intermediate free radicals, ERS spectroscopy may be the best method, nevertheless by this technique the ·OH radicals in the solution cannot be directly detected. Therefore, indirect techniques such as spin trapping with 5,5-dimethyl-1-pyrroline-N-oxide (DMPO) can be utilized. Under the conditions usually employed, the DMPO/·OH adduct can originate from the reaction of a precursor DMPO/superoxide adduct or from the oxidation of DMPO itself followed by a reaction with water.

Although the reactivity of iron complexes is recognized, the various complexes have different reaction rates for secondary reactions. Such reactions involve the iron complexes with hydroxyl radical trapping adducts under conditions normally employed. For example, the Fe^{III}DTPA (diethylenetriaminepentaacetic acid) complex can be reduced by oxidation of the DMPO/·OH spin adduct. These reactions, if not recognized, can lead to confounding stoichiometric calculations and kinetic comparisons in Fenton's reaction.

When the radicals formed from the trapped molecules present an additional reaction, these trapping experiments are also difficult to interpret. Thus, studies conducted by Lloyd *et al.*, 1997 [10], concluded that the nature of the chelating agent is an important factor, for example in the Fenton reaction $Fe^{II}EDTA$, $Fe^{II}DTPA$ and $Fe^{II}HEDTA$ with trapping alcohols. They found that in the studies realized with $Fe^{II}HEDTA$, the intermediate formed had better oxidizing properties than the hydroxyl radical, and that in the other two cases, the properties of the intermediate were equivalent to those of the hydroxyl radical. $Fe^{II}DTPA$ reacted with H_2O_2 to produce oxidizing species whose properties in *tert*-butyl alcohol trapping experiments did not match those of the ·OH radicals (reaction 77).

$$Fe^{II}(DTPA)^{3-} + H_2O_2 \rightarrow [Fe^{IV}(DTPA)OH]^{2-} + OH^- \tag{77}$$

Furthermore, kinetic results different from hydroxyl radicals can also be obtained when radicals formed from alcohols in the Fenton reaction react with iron (II) and (III).

It is possible that the oxygen atom was originated from hydrogen peroxide in both the ferric intermediate and the hydroxyl radical, although this has not been demonstrated. In addition, an exchange of oxygen atoms would occur if the hydroxyl radicals reacted with water. One possibility is that iron-bound hydroxyl radical species may exchange oxygen with water; however, in Fenton's reaction it has been shown that there is no exchange of oxygen atoms between ^{17}O-labeled H_2O_2 and the water solvent and *vice versa*.

The trapped hydroxyl radicals were derived exclusively from hydrogen peroxide and not from water, it was also concluded since, according to the hydroxyl radical generation method utilized, the percentages of ^{17}O-labeled hydroxyl radicals trapped by DMPO and from the original hydrogen peroxide were the same. Similarly, the Fenton reaction with standard H_2O_2, and with ^{17}O-labeled water, show that in the complementary reaction, none of the hydroxyl radicals were derived from the solvent water. In either case, the DMPO trapping was fast enough to avoid significant randomization of hydroxyl radicals with the water solvent.

Hydroxyl radicals are produced by homolytic cleavage from the photolysis of hydrogen peroxide. When comparing the photolysis results with the radicals from the Fenton reaction, it is observed that in the latter reaction, the labeled radical was also trapped before an isotopic exchange could take place. Reactions 76 and 77 implies that a fraction of hydroxyl radicals formed is bound to the iron complex. If this occurs, then the dissociation of radicals in ^{17}O-labeled species may not compete with intramolecular electron transfer in which water molecule bonds are involved, resulting in isotope randomization.

The above, does not exclude direct reaction of the ferryl intermediate $[Fe=O]^{2+}$ with DMPO to form DMPO/·OH when the ferryl oxygen is derived from H_2O_2 and is not exchanged with water binding. However, in some iron complex, oxygen exchange with hydroxyl radicals of similar reactivity, or such species dissociating into hydroxyl radicals, is excluded. Similarly, in DMPO/·OH generation, the possibility of an intermediate DMPO radical cation reacting with the hydroxyl anion to form DMPO/·OH by the Fenton reaction or by photolysis of H_2O_2 is excluded.

1.3. CYTOTOXICITY OF REACTIVE OXYGEN SPECIES (ROS)

Oxidative stress involves structural and functional changes produced by an unbalance between the generation of cell-damaging molecules and the cell's ability to detoxify itself. This imbalance occurs when unfavorable interactions of molecular oxygen (O_2) or its reactives (ROS) with biomolecules take place. Among ROS, the most abundant are: superoxide (O^-_2), hydroxyl radical (·OH), peroxynitrite ($ONOO^-$) and hydrogen peroxide (H_2O_2).

1.3.1. Superoxide

The O^-_2 superoxide anion radical is produced by the one-electron reduction of the oxygen triplet through an enzymatically catalyzed reaction, mainly in biological systems. Usually, this reaction occurs in mitochondria and microsome membranes. Once O^-_2 is generated, it can act as a free radical, an electron oxidizer, an electron reductant, or a weak nucleophile. Radical generation normally occurs by the release of hydrogen atoms added to the double bonds.

However, for O^-_2, these reactions are slower. In the presence of transition metals or quinones the O^-_2 donates electrons to the peroxides. Conversely, O^-_2 species can also act as oxidants on an electron, thus converting hydroquinone to semiquinone radicals with subsequent H_2O_2 production [11] (Fig. **2**).

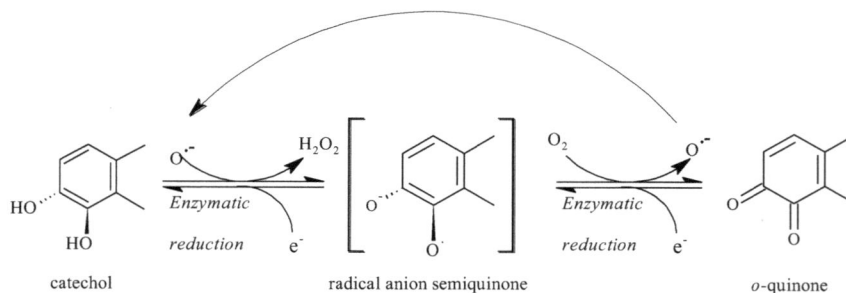

Fig. (2). Catechol-quinone redox cycle and superoxide dismutation.

Microsomal membranes, which include the membranes of lysosomes, peroxisomes, and the endoplasmic reticulum, are abundant in enzymes responsible for the detoxification of xenobiotics or the oxidation of fatty acids and steroids.

Microsomes contain mainly two different electron-carrying protein complexes and their isoforms: 1) NADH cytochrome b_5 reductase/cytochrome b_5 that contributes to the fatty acid acyl coenzyme A desaturase system and 2) NADPH-dependent cytochrome P450 reductase/ cytochrome P450- dependent mono oxygenase (CYP) complex that participates in the oxidation of endogenous and exogenous substrates by O_2. During these reactions, more NAD(P)H is oxidized, and more oxygen is consumed than normally needed. Excess O_2 can lead to O_2^{-} production.

The increase in ROS levels caused by UV radiation in human keratinocytes is mediated by the protein complex of the Nox family. Nox proteins are NADPH-dependent oxidases found in the plasma membranes of epithelial cells of the skin and colon. They are also found in endothelial cells, vascular smooth muscle cells, cardiac myocytes, fibroblasts, and thyroid cells.

Depending on the degree of activation and the cytotoxic signal, Nox 1 activation induces necrotic cell death through elevated ROS production. They also interfere with cell differentiation, thus contributing to epithelial cell transformation. Nox class enzymes are mainly expressed in phagocytic cells such as neutrophils and macrophages in contrast to NADPH-dependent oxidases. These cells produce large amounts of O_2^{-} (respiratory or oxidative burst) for defense against small solid particles, certain chemicals, and microorganisms. Oxidative burts facilitate tissue inflammation and releases cancer-promoting cytokines.

Electron transport chain located in the mitochondrial inner membrane containing flavoproteins, non-*heme* iron sulfur proteins and iron- and copper-containing cytochromes (b_{562}, b_{566}, c_1, c, a, c_3) and coenzyme Q (ubiquinone-ubiquinol), oxidize the cofactors NADH and $FADH_2$. There also free radicals are constantly formed of which O_2^{-} formation is proportional to mitochondrial O_2 utilization. The levels of O_2^{-} formation correspond to 1-2% of total oxygen consumed but when electron flow is inhibited (by antimicin or rotenone), it increases considerably. There are at least two sites of O_2^{-} production: the flavoprotein complex NADH dehydrogenase and the ubiquinone segment cytochrome b_{562}/b_{566}, respectively. Under physiological conditions, cytochrome b_{566} is the most important, but when mitochondrial membrane damage occurs, it can be substituted by ubisemiquinone due to elevated proton concentrations. Normally, low amounts of O_2^{-} are released from mitochondria due to high levels of matrix-associated manganese-dependent superoxide dismutase (MnSOD) [12].

1.3.2. Hydrogen Peroxide

The toxicity of hydrogen peroxide is indirect since by itself it is not reactive, but through the Fenton reaction, it oxidizes Fe(II) to Fe(III) to generate hydroxyl radicals ·OH [13]. Peroxisome metabolism, O^-_2 dismutation by SOD and autoxidation of compounds during the redox cycling of the electron transport chain of both the microsome and mitochondria are the main cellular sources of hydrogen peroxide. Like O^-_2 and due to its stability, H_2O_2 generated in certain compartments may represent a danger to biomolecules at distant sites. H_2O_2 is mainly removed by thioredoxin-dependent peroxidases (peroxiredoxins), glutathione peroxidases (GPx) and peroxisomal catalase.

1.3.3. Hydroxyl Radical

The hydroxyl radical ·OH is the most toxic reactive species of the ROS. It has a very short half-life (~1ns), and interacts with biomolecules once generated. In the Fenton reaction, when copper or iron ions are present, electron donors such as NADPH, hydroquinones, ascorbic acid, catechins, or glutathione (GSH) facilitate the formation of ·OH radicals from hydrogen peroxide H_2O_2. Low molecular weight iron complexes contribute mainly to the formation of high levels of ·OH. Due to its ability to generate ROS iron can exert strong cytotoxicity and contribute to lipid peroxidation (LPO), if not controlled through binding to specific iron storage proteins such as ferritin. It is known that certain xenobiotic compounds, such as catechols, trigger the release of iron from ferritin and thus initiate LPO in exposed cells.

1.3.4. ROS Damage to Macromolecules

Human cells function in a reduced state; however, a certain degree of localized oxidation is needed. For example: for proteins to fold properly and disulfide bridges to be allowed to form, it is important that they are in a more oxidized state than the rest of the cell. Despite the above, when there is an overproduction of oxygen free radicals or when defense systems are damaged, these highly reactive chemical species are harmful.

Interactions of oxygen free radicals with cellular constituents lead to alterations in cellular metabolism and cause subcellular damage that can lead to the onset of disease and even death. The intermediates and reactive oxygen by-products derived from oxidative metabolism are the main threat to the homeostasis of aerobic organisms.

Oxygen free radicals, including molecular oxygen derivatives, can interact with almost any of the biomolecules that are constituents of cells. Proteins, lipids, and carbohydrates are fundamental targets of oxygen free radical reactions (Fig. **3**); however, other cellular components may also be sensitive to the effects of these potent oxidants, among these components are neurotransmitters such as serotonin or adrenaline, various enzymatic cofactors, antioxidants, aromatic amino acids and sulfur [8].

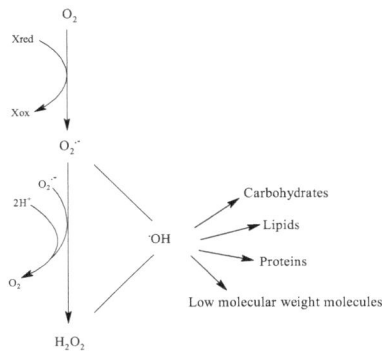

Fig. (3). Free radical production and its relationship to cellular components.

1.3.4.1. Proteins

Protein susceptibility to free radical damage depends on amino acid composition. Because of free-radical reactivity to molecules with double bonds or containing sulphur groups, proteins with a large proportion of amino acids such as tryptophan, tyrosine, phenylalanine, hystidine, methionine and cysteine can easily be attacked by free radicals. In any case, the magnitude of oxidative damage will depend on whether these amino acids are part of functional groups responsible for the activity and/or conformation of these proteins. In this respect, it has been established that enzymes such as papain or glyceraldehyde-3-phosphate dehydrogenase, whose catalytic activity depends on these amino acids, in the presence of free radicals are inhibited. α_1-antiprotease is also inactivated when methionine in its active center is oxidized to sulfoxide.

Free-radical reactions with proteins also result in protein structural alterations, which cause cross-links and aggregation phenomena that may be mediated by intramolecular and intermolecular disulfide bridges. When a compound containing a sulfhydryl (RHS) group is oxidized by a free radical, a thiyl (RS·) radical is formed that can interact with another to form a disulfide (RSSR) bridge (reactions 78 and 79). These reactions could explain the protective effect of compounds containing sulfhydryl groups against free-radical damage.

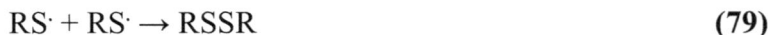

$$RSH + A^. \rightarrow RS^. + AH \tag{78}$$

$$RS^. + RS^. \rightarrow RSSR \tag{79}$$

Peptide bonds and amino acids such as proline or lysine, which are normally more resistant to modification, by the action of some highly reactive oxygen species may also be affected. Proline residue oxidation, mediated by hydroxyl or superoxide radicals, followed by hydrolysis of peptide bonds, is the proposed mechanism to explain oxidative excision and protein deamination.

In addition to oxidizing amino acids, oxygen free radicals can react directly with the metal bonds of many metal proteins, modifying their redox state. For example, iron from haemoglobin or catalase may react with the superoxide radical and become its inactive part Fe^{3+}. CuZn superoxide dismutase can react with hydrogen peroxide to generate hydroxyl radical, which can attack a residue of hystidine from the enzyme's active center. Finally, oxygen free radical reactions with proteins can also generate by-products that can extend the initial damage. For example, *N*-formylquinurenine, which originates from tryptophan oxidation, may react with compounds containing amino groups and cause cross-links between lipids and/or proteins.

1.3.4.2. Lípids

Oxidative stress damages macromolecules such as proteins, structural carbohydrates, and lipids. Lipid peroxidation is particularly affected, because the formation of products in this process leads to the easy propagation of free radicals. Free radicals can react with the polyunsaturated fatty acids of membrane lipids, causing oxidative damage to the membranes. This phenomenon, known as lipid peroxidation, is the source, for example, of the alteration of the cover of low-density lipoproteins (LDL).

Lipid peroxidation can be initiated by hydroxyl and hydroperoxyl radicals, but not by superoxide radicals. The $O^._2$ may have a small role in the breakdown of the hydroperoxides formed (reaction 80).

$$O^._2 + ROOH \rightarrow O_2 + OH^- + RO^. \tag{80}$$

Lipid peroxidation consists of three phases: initiation, propagation, and termination. The initiation phase involves an extraction of hydrogen atoms. The first hydrogen atom can be extracted by the hydroxyl radical ($\cdot$OH), the alkoxyl radical (RO$\cdot$), the peroxyl radical (ROO$\cdot$) and possibly HO$\cdot_2$, but not by H_2O_2 or the superoxide radical O$\cdot_2$. Due to the removal of a hydrogen atom from the methylene group (-CH2-), the polyunsaturated fatty acids of membrane lipids are very sensitive to peroxidation, leaving an unpaired electron on the -$\cdot$CH- carbon. The presence of a double bond in the fatty acid weakens the C-H bond on the carbon atom adjacent to the double bond and thus facilitates the extraction of H$\cdot$.

Lipid peroxidation is initiated by the existence of an oxygen free radical (X$\cdot$), by light or by metal ions (Fig. **4**). In the initial reaction of $\cdot$OH with polyunsaturated fatty acids, an alkyl radical (R$\cdot$) is produced, which in turn reacts with molecular oxygen to form a peroxyl radical (ROO$\cdot$). The ROO$\cdot$ extracts a hydrogen from the adjacent fatty acid and produces a lipid hydroperoxide (ROOH) and a second alkyl radical. The ROOH presents a reductive fragmentation by reduced metals such as Fe^{2+}, thus originating a lipid alkoxyl radical (RO$\cdot$). By abstraction of hydrogen atoms, both alkoxyl and peroxyl radicals stimulate the chain reaction of lipid peroxidation [14].

ROOH may produce reactive aldehyde products such as 4-hydroxynonenal (HNE), acrolein and malondialdehyde (MDA), usually in the presence of metals or ascorbate. Through detoxification by glutathione-S-transferase and glutathione peroxidase, reduced glutathione (GSH) can be depleted due to lipid and aldehyde overproduction.

The structure of HNE contains three functional groups. The most important is a conjugated system of C=C double bonds and a C=O carbonyl group, which contributes the partially positive charge to carbon 3. On carbon 4, this partial positive charge is enhanced by the inducing effect of the hydroxyl group. The HNE is the main product of the oxidation of the ω-6 polyunsaturated lipid chain (linoleic and arachidonic), is the most reactive of these compounds and is an important mediator of free radical damage.

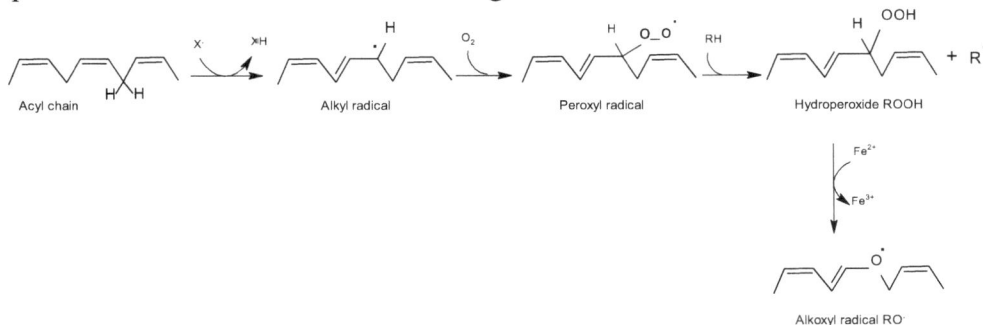

Fig. (4). Lipid peroxidation reactions.

Alterations in membrane fluidity and permeability, ion transport and inhibition of metabolic processes are some of the damages caused by lipid peroxidation. The ribosomes separate from the endoplasmic reticulum, the mitochondrial electronic transport deteriorates, and the mitochondria are lysed; the lysosomes are also lysed, and their enzyme content is discharged into the cytosol. The nuclear membrane also peroxidates, releases low molecular weight aldehydes that can inhibit protein synthesis. Damage to the mitochondria induced by lipid peroxidation may promote ROS regeneration [14].

The complete process can be represented as follows:

1) Initiation:

$$ROOH + Metal^{(n)+} \rightarrow ROO^{\cdot} + Metal^{(n-1)+} + H^{+} \tag{81}$$

$$X^{\cdot} + RH \rightarrow R^{\cdot} + XH \tag{82}$$

2) Propagation:

$$R^{\cdot} + O_2 \rightarrow ROO^{\cdot} \tag{83}$$

$$ROO^{\cdot} + RH \rightarrow ROOH + R^{\cdot}, etc. \tag{84}$$

3) Termination:

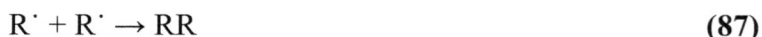

$$ROO^{\cdot} + ROO^{\cdot} \rightarrow ROOR + O_2 \tag{85}$$

$$ROO^{\cdot} + R^{\cdot} \rightarrow ROOR \tag{86}$$

$$R^{\cdot} + R^{\cdot} \rightarrow RR \tag{87}$$

1.3.4.3. Carbohydrates

The carbohydrates are also targeting the oxygen-free radicals. Monosaccharides such as glucose, mannitol, deoxysugars and certain nucleotides can easily react with hydroxyl radicals to produce new highly reactive free radicals [8]. The glycosylation of proteins, therefore, makes them more susceptible to free radical oxidation.

Free radicals can also react with carbohydrate polymers, usually inducing fragmentation. Hyaluronic acid is a glycosaminoglycan that is constituted by repeated units of glucuronic acid and *N*-acetylglycosamine and whose function is to maintain the synovial fluid viscosity of the joints. When hyaluronic acid is exposed to oxygen-free radical generating systems, it is depolymerized and consequently loses its lubricating properties.

CONCLUSION

Evolutionary theories exist about the time and events that led to the emergence of oxygen on earth and its accumulation in the terrestrial atmosphere. Similarly, there is scientific evidence supporting the evolutionary mechanisms that allowed the first life forms to incorporate oxygen into their metabolism by means of suitably adapted enzyme systems. Analyses of the behavior of electrons from aqueous environments in reactions with oxygen free radicals have been performed. When the accumulation of oxygen free radicals inside the cell exceeds the capacity of the cells to eliminate them, oxidative stress occurs, damaging macromolecules such as proteins, lipids, and carbohydrates.

REFERENCES

[1]   Dismukes, G.C.; Klimov, V.V.; Baranov, S.V.; Kozlov, Y.N.; DasGupta, J.; Tyryshkin, A. The origin of atmospheric oxygen on Earth: the innovation of oxygenic photosynthesis. *Proc. Natl. Acad. Sci. USA,* **2001**, *98*(5), 2170-2175.
[http://dx.doi.org/10.1073/pnas.061514798] [PMID: 11226211]

[2]   Holland, H.D. The oxygenation of the atmosphere and oceans. *Philos. Trans. R. Soc. Lond. B Biol. Sci.,* **2006**, *361*(1470), 903-915.
[http://dx.doi.org/10.1098/rstb.2006.1838] [PMID: 16754606]

[3]   Lyons, T.W.; Reinhard, C.T.; Planavsky, N.J. The rise of oxygen in Earth's early ocean and atmosphere. *Nature.* **2014**, *506*(7488), 307-315.
[http://dx.doi.org/10.1038/nature13068] [PMID: 24553238]

[4]   Brocks, J.J.; Logan, G.A.; Buick, R.; Summons, R.E. Archean molecular fossils and the early rise of eukaryotes. *Science,* **1999**, *285*(5430), 1033-1036.
[http://dx.doi.org/10.1126/science.285.5430.1033] [PMID: 10446042]

[5]   Gaillard, F.; Scaillet, B.; Arndt, N.T. Atmospheric oxygenation caused by a change in volcanic degassing pressure. *Nature,* **2011**, *478*(7368), 229-232.
[http://dx.doi.org/10.1038/nature10460] [PMID: 21993759]

[6]   Falkowski, P.G.; Godfrey, L.V. Electrons, life and the evolution of Earth's oxygen cycle. *Philos. Trans. R. Soc. Lond B Biol. Sci.,* **2008**, *363*(1504), 2705-2716.
[http://dx.doi.org/10.1098/rstb.2008.0054] [PMID: 18487127]

[7]   Ananyev, G.M.; Zaltsman, L.; Vasko, C.; Dismukes, G.C. The inorganic biochemistry of photosynthetic oxygen evolution/water oxidation. *Biochim. Biophys. Acta,* **2001**, *1503*(1-2), 52-68.
[http://dx.doi.org/10.1016/S0005-2728(00)00215-2] [PMID: 11115624]

[8]   Martínez, C.M. M. Estrés oxidativo y mecanismo de defensa antioxidante. In: *Cap. 18. En: GIL H. A. Tratado de nutrición*; Editorial Panamericana, **2010**; pp. 455-80.

[9]   Pálfi, T.; Wojnárovits, L.; Takács, E. Calculated and measured transient product yields in pulse radiolysis of aqueous solutions: Concentration dependence. *Radiat. Phys. Chem.,* **2010**, *79*(11), 1154-1158.
[http://dx.doi.org/10.1016/j.radphyschem.2010.06.004]

[10]   Lloyd, R.V.; Hanna, P.M.; Mason, R.P. The origin of the hydroxyl radical oxygen in the Fenton reaction. *Free Radic. Biol. Med.,* **1997**, *22*(5), 885-888.
[http://dx.doi.org/10.1016/S0891-5849(96)00432-7] [PMID: 9119257]

[11]   Goetz, M.E.; Luch, A. Reactive species: a cell damaging rout assisting to chemical carcinogens. *Cancer Lett.,* **2008**, *266*(1), 73-83.

[http://dx.doi.org/10.1016/j.canlet.2008.02.035] [PMID: 18367325]

[12]   Echtay, K.S. Mitochondrial uncoupling proteins--what is their physiological role? *Free Radic. Biol. Med.,* **2007**, *43*(10), 1351-1371.
[http://dx.doi.org/10.1016/j.freeradbiomed.2007.08.011] [PMID: 17936181]

[13]   Griendling, K.K.; Touyz, R.M.; Zweier, J.L.; Dikalov, S.; Chilian, W.; Chen, Y-R.; Harrison, D.G.; Bhatnagar, A. Measurement of reactive oxygen species, reactive nitrogen species, and redox-dependent signaling in the cardiovascular system: A scientific statement from the american heart association. *Circ. Res.,* **2016**, *119*(5), e39-e75.
[http://dx.doi.org/10.1161/RES.0000000000000110] [PMID: 27418630]

[14]   Murray, R.K.; Granner, D.K.; Mayes, P.A.; Rodwell, V.W. Harper's Ilustrated Biochemistry. In: *Twenty-sixth editions*; , **2003**. Lange Medical Books/ McGraw-Hill.

CHAPTER 2

Biological Oxidation

Abstract: Biological oxidation is the mechanism by which oxygen is supplied to living organisms and energy is produced. These processes involve the participation of oxygen incorporated into the substrates by oxidoreductase enzymes. The mechanisms of reaction of enzymes involved in both oxidation-reduction processes and electron transport chain mitcchondrial-ETC, and their relationship with reactive oxygen species (ROS) production, are presented here.

Keywords: Dehydrogenases, Hydroperoxidases, Mitochondrial Electron Transport Chain-ETC, Oxidases, Oxidoreductases, Oxygenases, Redox potential, ROS.

INTRODUCTION

The **oxidation** is a chemical process in which electron elimination occurs and **reduction** is the process by which electron gain occurs, so that oxidation is accompanied by the reduction of an electron. This oxidation-reduction principle allows us to understand the complexity of the biochemical nature of the metabolism of living organisms. However, some biological oxidations occur without the participation of molecular oxygen, such as hydrogenations. Earth life depends mainly on the supply of oxygen to carry out respiration processes, especially in higher organisms where cells produce energy in the form of ATP with water formation from oxygen-hydrogen reactions. Oxygen is incorporated into substrates by means of enzymes called *oxygenases*.

1. REDOX POTENTIAL

In oxido-reduction reactions, the free energy changes are related to the ability to donate or accept electrons by the reactants. The free energy change is expressed as $\Delta G^{0\prime}$ but can also be expressed as oxide-reduction potential or redox potential (E'_0). For biochemical systems, the redox potential (E'_0) is expressed at pH 7.0 where the hydrogen electrode potential is -0.42 volts, considering an E_0 systems redox potential referred to the hydrogen electrode potential (0.0 volts at pH 0.0).

Biological oxidation-reduction processes involve enzymes called **oxido reductases** and are mainly classified into four groups: **oxidases**, **dehydrogenases**, **hydroperoxidases** and **oxygenases** [1].

1.1. Oxidases

Utilizing oxygen as a hydrogen acceptor, oxidases catalyze the removal of hydrogen from a substrate. Water or hydrogen peroxide is formed as a product of the reaction (Fig. **1**).

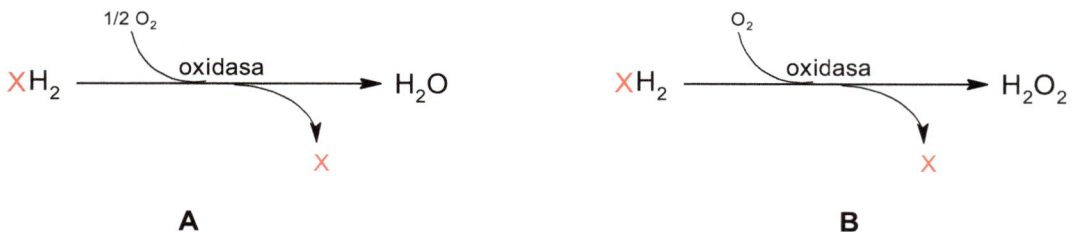

Fig. (1). Oxidase activity in metabolite oxidation **A.** Forming H_2O. **B.** Forming H_2O_2.

Cytochrome oxidase is found in hemoglobin, myoglobin, and other cytochromes. It is an oxidase containing copper and the typical *heme* prosthetic group. It is widely distributed in tissues and is the terminal component in the electron transport chain of mitochondria. The substrate is oxidized by hydrogenases and the electrons produced are transported by cytochrome oxidase to oxygen, which is the final electron acceptor. Carbon monoxide, hydrogen sulfide and cyanide inhibit the enzyme. This enzyme is also known as *cytochrome a_3*.

The following reaction is catalyzed by cytochrome *c* oxidase in solution:

$$4\text{ferrocytochrome } c + 4H^+ + O_2 \rightarrow 2H_2O + 4\text{ferrycytochrome } c \qquad (1)$$

For reaction 1, the average force is about 800-250 = 550 meV or ~12.7 kcal/mol per electron, *i.e.,* ~51 kcal/mol for the complete four-electron reaction [2]. The free energy exchange per electron for this reaction is 565 meV (~13 kcal/mol) and involves the release of 1 oxygen atom, which at 25°C corresponds to 1.2 mM O_2. Being a higher concentration than the O_2 concentration in mammalian tissues (0.005 - 0.025mM) and of the water-saturated atmosphere (0.258 mM at 25°C).

For the electron acceptor pair O_2/H_2O, a potential of 815 mV at pH = 7 ($E_{m.7}$) was determined. Cytochrome *c* oxidase is the 4-electron donor and the *c*-type cytochrome of some bacteria and mammalian cytochrome *c* has a potential of 250 mV at $E_{m.7}$.

Flavoproteins are oxidases that contain as their prosthetic group flavin mononucleotide (FMN) or flavin adenine dinucleotide (FAD). FMN and FAD, which are non-covalently bound to their respective apoenzymes, are formed in the organism from the vitamin riboflavin. Many riboflavin dehydrogenases are associated with electron transport or the respiratory chain. The metaloflavoproteins contain as cofactors one or more metals.

Flavoproteins include **xanthine oxidase** is important for the conversion of purine bases into uric acid, its structure contains molybdenum; an NMF-enzyme that has specificity for the oxidative deamination of L-amino acids is **L-amino acid oxidase** and is found in the kidneys; **aldehyde dehydrogenase** is a FAD enzyme that acts on aldehyde and *N*-heterocyclic sulfates, contains non-*heme* iron, molybdenum and is found in mammalian liver. By means of a two-step reaction, the oxidation and reduction processes of these enzymes are carried out. This mechanism is also present in ubiquinone, an electron transporting coenzyme (Fig. **2**).

Fig. (2). Mechanism of oxidoreduction of ubiquinone *via* semiquinone-free radical.

Dehydrogenase enzymes participate in hydrogen transfer in coupled oxidoreduction reactions, where hydrogenases utilize coenzymes or hydrogen transporters such as NAD^+, despite having substrate specificity. Considering that the reactions are reversible, then in the cell the equivalent reductants are freely transferred. Such reactions are useful for oxidative processes to occur in the absence of oxygen, as occurs in the glycolysis anaerobic phase where the substrate is oxidized at the expense of another.

Furthermore, hydrogenases are also components of the electron transport chain from substrate to oxygen (Fig. **3**).

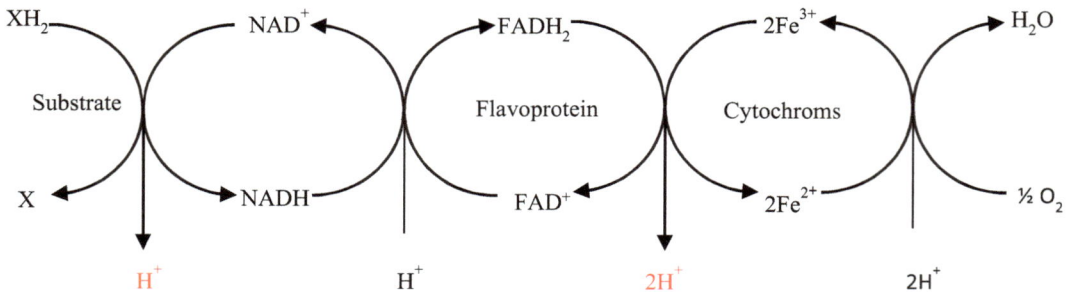

Fig. (3). Transport of reducer equivalents through the respiratory chain.

Some dehydrogenases are formed in the body from vitamin niacin and utilize NAD^+ or $NADP^+$ or both. Specific substrates of dehydrogenases reduce coenzymes that are also reoxidized by an electron acceptor available (reaction 2).

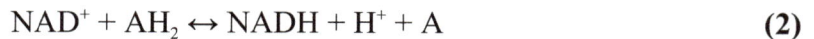

$$NAD^+ + AH_2 \leftrightarrow NADH + H^+ + A \qquad (2)$$

In the oxido-reduction mechanism of nicotinamide coenzymes when reduced by an AH_2 substrate, a stereospecificity occurs near position 4 of the coenzyme. According to the specificity of the reaction catalyzed by the dehydrogenase, a hydrogen atom is removed from the substrate as a hydride ion, H^- (one atom with two electrons) and transferred to position 4 where it binds to position A or B (Fig. 4).

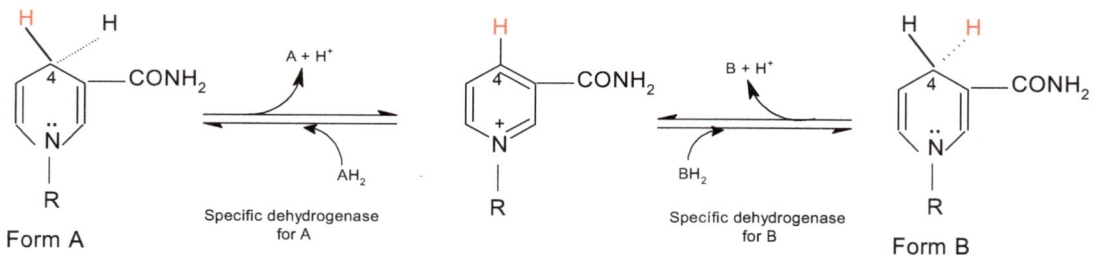

Fig. (4). Oxide-reduction mechanism of nicotinamide coenzymes.

They can dissociate from their specific apoenzymes in a free and reversible form. The oxidation-reduction reactions of oxidative metabolism mainly as glycolysis, citric acid cycle and electron transport chain of mitochondria, are catalyzed by NAD-dehydrogenases.

These enzymes are involved in pentose phosphate pathway reductive synthesis and steroid and fatty acid synthesis through extramitochondrial pathways [1].

1.2. NADH Dehydrogenase

Electron transport between NADH and high potential redox compounds is performed by NADH dehydrogenase enzyme (Fig. **3**). However, dehydrogenases such as acyl-CoA dehydrogenase, succinate dehydrogenase and glycerol-3-phos phate dehydrogenase transfer reducing equivalents from the substrate to the respiratory chain directly (Fig. **4**). In addition, by dihydrolipoyl dehydrogenase, flavin dehydrogenase performs dehydrogenation of reduced lipoate which is an intermediate of oxidative carboxylation of pyruvate and α-ketoglutarate. The function of transporter between acyl-CoA dehydrogenase and the respiratory chain is performed by electron transfer flavoprotein (Fig. **5**).

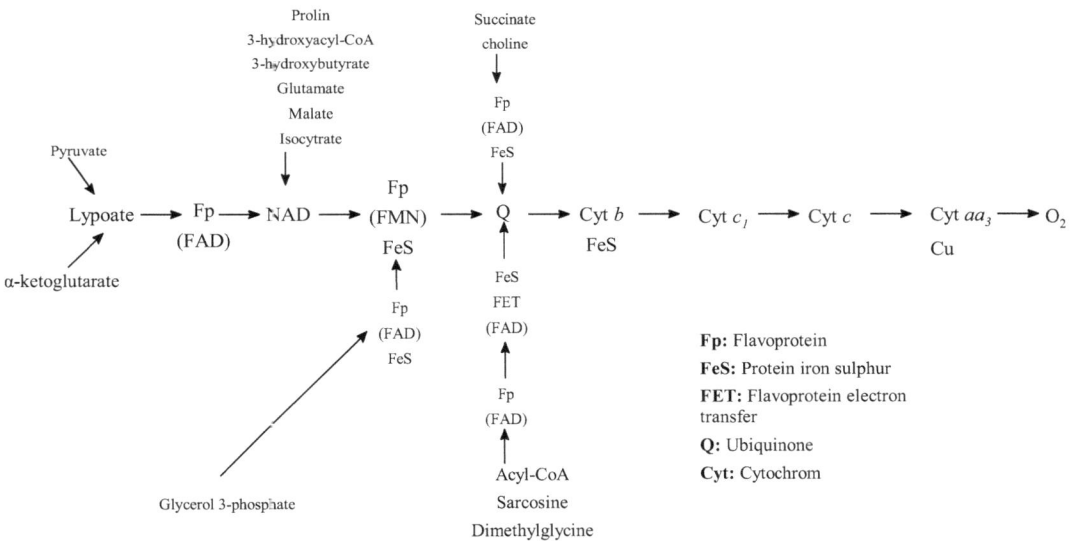

Fig. (5). Mitochondrial respiratory chain components.

Cytochromes are considered dehydrogenases except for cytochrome oxidase. They are iron-containing hemoproteins that during oxidation and reduction oscillate between Fe^{3+} and Fe^{2+}. They participate in the respiratory chain as electron transporters from the flavoproteins on the cytochrome oxidase side. Cytochromes a, a_3, b, c and c_1 are present in the respiratory chain. They are also found in bacteria, yeast, endoplasmic reticulum -cytochrome P450 and -b_5 from animal cells, and in plant cells.

1.3. Hydroperoxidases

The hydroperoxidases include the enzymes peroxidases and catalases. They protect the organism against dangerous peroxides, are found in the blood tissue and are involved in the metabolism of eicosanoids. Hydrogen peroxide and organic peroxide are the substrates of these enzymes, where the proto*heme* group is the prosthetic group. Hydrogen peroxide is reduced at the expense of electron acceptors such as cytochrome *c*, ascorbate and quinones, in the reaction catalyzed by peroxidases, whose general reaction is as follows:

$$PEROXIDASE$$
$$H_2O_2 + AH_2 \longrightarrow 2H_2O + A \qquad (3)$$

The removal of H_2O_2 hydrogen peroxide and lipid hydroperoxide is carried out by the enzyme **glutathione peroxidase** through the reduction of glutathione. This enzyme is found in erythrocytes and tissues containing selenium as a prosthetic group. Protecting against peroxide oxidation of membrane lipids and hemoglobin.

Catalase is a hemoprotein containing four *heme* groups. This enzyme utilizes hydrogen peroxide both as a donor and as an electron acceptor. Thus, its peroxidase activity allows it to utilize one H_2O_2 molecule as electron donor and another H_2O_2 as electron acceptor:

$$CATALASE$$
$$2H_2O_2 \longrightarrow 2H_2O + O_2 \qquad (4)$$

Alfonso-Prieto *et al.*, 2009 [3] analyzing the molecular mechanisms of catalase activity, they concluded that the electron transfer occurs in a two-step reaction, each time transferring one electron since the two electrons are not transferred simultaneously. Initially, the enzyme is oxidized to an iron intermediate, compound I (CpdI). It is reduced back to the resting state by further reacting with H_2O_2.

Under *in vivo* conditions, it has been demonstrated that the protection of organisms against H_2O_2 increases in the presence of catalase-peroxidase. The enzyme catalase is found in tissues such as blood, bone marrow, bone, kidney, liver, and mucous membranes. Their main function is to eliminate the hydrogen peroxide that is produced by the enzyme oxidase.

Peroxisomes, found in the liver and other tissues, are an important source of oxidases and catalases. Thus, H_2O_2-producing enzymes are present in the same location as H_2O_2-removing enzymes. Additional sources of H_2O_2 are xanthine oxidase and the mitochondrial and microsomal electron transport systems.

1.4. Oxygenases

Both the synthesis and degradation of metabolites involve oxygenase enzymes. Oxygenases mechanism of action is through the incorporation of oxygen into the substrate in a two-step reaction: 1) at the active site of the enzyme, oxygen is bound and 2) oxygen is reduced or transferred to the substrate. Oxygenases are divided into deoxygenases which incorporate both oxygen atoms into the substrate. The general reaction is:

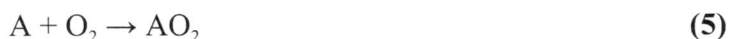

$$A + O_2 \rightarrow AO_2 \tag{5}$$

Iron-containing liver enzymes such as 3-hydroxyanthranilate dioxygenase, homogentisate deoxygenase and the enzyme L-tryptophan dioxygenase which utilizes the *heme* group are examples of oxidase enzymes.

For their part, monooxygenases act as oxidases and hydrolases. Their mechanism of action incorporates only one oxygen atom to the substrate while the other oxygen atom is reduced to water, being necessary the presence of an additional electron donor or cosubstrate (z):

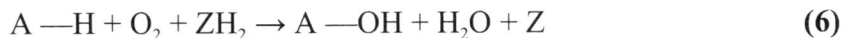

$$A\!-\!H + O_2 + ZH_2 \rightarrow A\!-\!OH + H_2O + Z \tag{6}$$

Cytochrome P450 is an important superfamily of monoxygenases, and there are more than 1000 enzymes of this class, which contain the *heme* group. For the reduction of these cytochromes both NADH and NADPH donate reducing equivalents and the cytochromes are also oxidized by substrates. Thus, in the mechanism of P450 the redox pair protein interaction reactions and the consumption of reducing equivalents such as NADPH are involved. In the hydroxylase cycle the series of enzymatic reactions involved in the oxidation and reduction mechanisms of cytochromes P450 are presented (Fig. **6**).

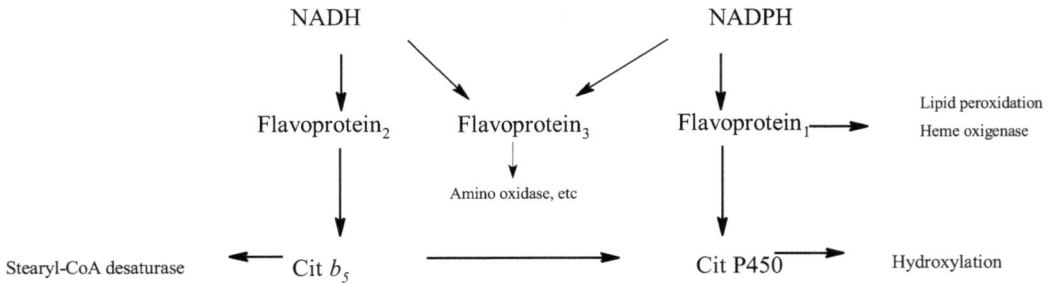

Fig. (6). Electron transport chain in microsomes.

Denisov *et al.*, 2005 [4] argue that the catalytic mechanism of P450 includes: (1) the binding of oxygen to the reduced *heme* iron and the formation of an oxygenated *heme* iron $Fe^{2+}-OO$ or $Fe^{3+}-OO^-$, (2) reduction of an electron from this complex to a ferric peroxide state $Fe^{2+}-OO^{2-}$, which is protonated to form the hydroperoxide $Fe^{3+}-OOH^-$, (3) a second protonation of the $Fe^{3+}-OOH^-$ complex in the distal oxygen atom to form a transient unstable $Fe-OOH_2$, followed by a heterolytic excision of the $O-O$ link and the release of the water molecule,   and (4) the various the reactions of porphyrin.

Electron transfer to O_2 generates the O^-_2 superoxide free radical anion, which produces free radical chain reactions, producing a destructive effect. The superoxide radical is formed in all tissues from the oxygen present in them. Superoxide dismutase (SOD) is the enzyme responsible for eliminating superoxide radicals in all aerobic organisms, although not in anaerobes, protecting them from oxygen-free radical toxicity.

When the reduced flavins such as xanthine oxidase are univalently re-oxidized by molecular oxygen, the superoxide radical is produced:

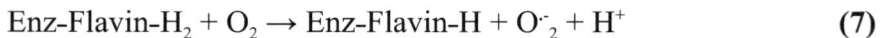

$$\text{Enz-Flavin-H}_2 + O_2 \rightarrow \text{Enz-Flavin-H} + O^-_2 + H^+ \qquad (7)$$

Superoxide can be eliminated by superoxide dismutase (SOD), in a reaction

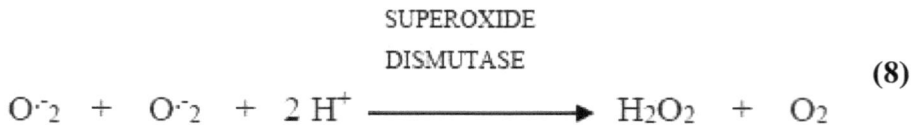

$$O^-_2 + O^-_2 + 2H^+ \xrightarrow{\quad \text{SUPEROXIDE DISMUTASE} \quad} H_2O_2 + O_2 \qquad (8)$$

where it acts as both an oxidant and a reductant. It may also reduce cytochrome *c*. This is the main mechanism by which SOD protects organisms from the toxic effects of superoxide. The enzyme is present mainly in mitochondria and cytosol of aerobic tissues.

2. MITOCHONDRIAL ELECTRON TRANSPORT CHAIN

The main source of reactive oxygen species (ROS) in most eukaryotic cells is mitochondrial energy metabolism [5]. Mitochondrial ROS are known to be involved in signaling processes and degenerative events constituting important determinants of cell function. During normal cell metabolism, they are formed mainly in the cytosol, plasma membrane, mitochondria, and peroxisomes and by radiolysis of water.

Endogenous antioxidants such as catalase, glutathione peroxidase, and thiol-based scavengers such as glutathione can minimize the damage caused by excessive production of free oxygen radicals to membranes, genes, and enzymes. However, they cause oxidative stress. Despite this, the cell can survive the confinement of oxygen free radicals in the compartments, the toxicity of small amounts of ROS, the catalytic action of enzymes such as SOD and the action of reducing agents such as GSH cysteine, tocopherol-vitamin E, ubiquinol and ascorbate, among others.

In the mitochondria, due to the monoelectronic reduction of O_2 (reaction 9), the first reactive oxygen-ROS species is produced, which is the $O_2^{\cdot-}$ superoxide ion.

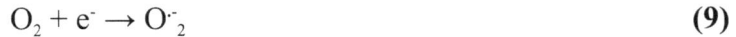

$$O_2 + e^- \rightarrow O_2^{\cdot-} \tag{9}$$

A small amount of $O_2^{\cdot-}$ is produced when in the mitochondrial electron transport chain most of the consumed oxygen is reduced in a four-electron reduction. Through the Mn-SOD activity of the mitochondrial matrix (Fig. 7) and Cu, Zn-SOD in the intermembrane space, $O_2^{\cdot-}$ is transformed into a more stable species, H_2O_2 (reaction 10).

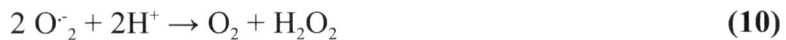

$$2\,O_2^{\cdot-} + 2H^+ \rightarrow O_2 + H_2O_2 \tag{10}$$

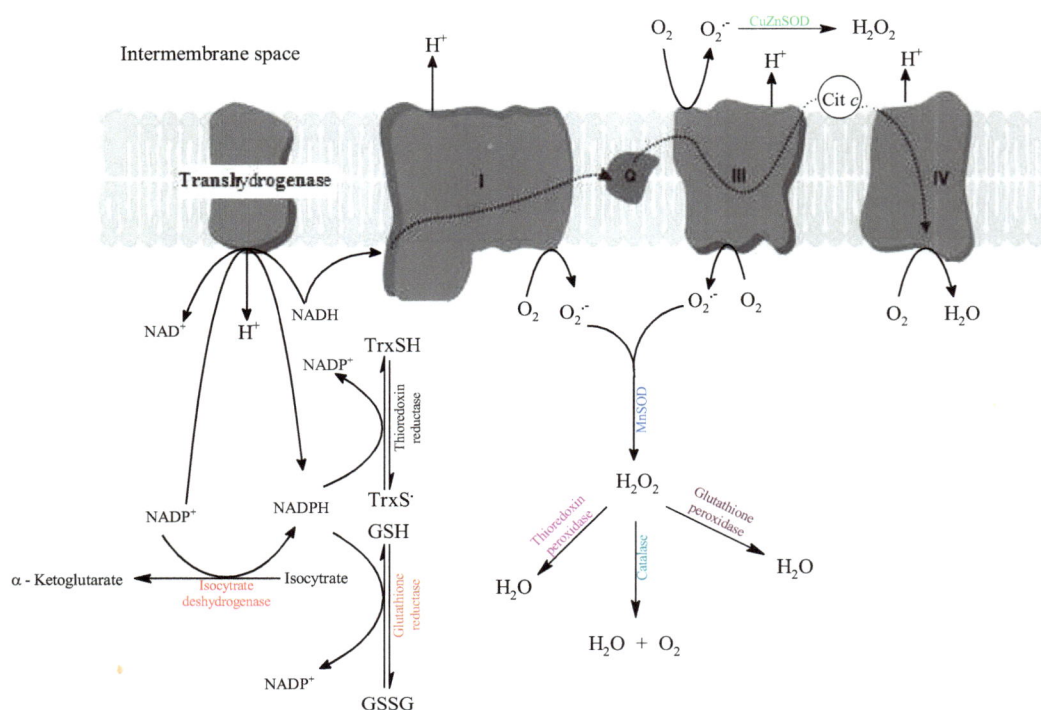

Fig. (7). ROS mitochondrial metabolism. Explanation in the text.

The H_2O_2 produced in mitochondria acts as a signaling molecule in the cytosol, in cell cycle energy metabolism, in cellular redox balance and as a response to stress. However, as H_2O_2 is relatively stable, it permeates the membrane and diffuses into the cell where it is removed by cytosolic antioxidant enzymes such as glutathione peroxidase, catalase and thioredoxin peroxidase.

To remove the H_2O_2 formed, mitochondria possess peroxiredoxins and reductases, such as the thioredoxin peroxidase/thioredoxin reductase system, which utilizes the electrons yielded by thioredoxin, and the glutathione peroxidase/glutathione reductase system, which utilizes GSH as a source of electrons to remove H_2O_2. In turn, mitochondrial glutathione and thioredoxin are reduced by NADPH. Thus, mitochondrial antioxidant capacity is related to NADPH levels.

To promote NADPH formation, $NADP^+$ is reduced in mitochondria through $NADH/NADP^+$ transhydrogenase activity, which acts as a proton pump utilizing the electrochemical H^+ gradient from respiration. Thus, mitochondrial coupling and membrane potential is coupled to the redox potential. However, low rates of H_2O_2 removal, can result in oxidative damage when mitochondria are not fully coupled or if the potential membrane decreases, such that transhydrogenase-bound energy cannot respond rapidly to high levels of NADPH oxidation. Through isocitrate dehydrogenase activity, $NADP^+$ can be maintained in a reduced state (Fig. 7).

Mitochondrial complex III (Q-cytochrome c oxidoreductase) is a well-documented source of mitochondrial ROS. This respiratory complex receives electrons from the reduced coenzyme Q (UQH_2) and donates them to cytochrome c. UQH_2 is oxidized to UQ in a complex of reactions involving first the formation of the semiquinone radical (UQ^-) at the Qp site of complex III, (oriented towards the intermembrane space) by electron donation from UQH_2 to the Rieske protein (ISP iron-sulfurized protein) and then to cytochrome c. An electron from the radical UQ^- formed at the Qp site is then transferred to the Qn site (oriented towards the mitochondrial matrix) where UQ is reduced to UQ^-. The Qn UQ^- is reduced to UQH_2 by an electron supplied by a second UQ^- formed on the Qp site [5].

On the sites, Qp and Qn is formed UQ^- as a result of this cycle. Since O_2 has access to one of the sites of the UQ^-/UQ complex, O^-_2 may be formed by electron donation from the highly reduced UQ^-/UQ pair. Alternatively, O^-_2 production can also occur in complex I (NADH-Q oxidoreductase) of the ubiquinone cycle, mainly at two sites: at the upstream site, which can be the FMN group or the iron-sulfurized centers or at the downstream site at the ubiquinone binding site. Through the metal-catalyzed Fenton reaction (reaction 10 chapter 1) ·OH hydroxyl radicals are produced from H_2O_2 when it is not metabolized by mitochondrial antioxidant systems. The ·OH hydroxyl radical is highly reactive. Therefore mitochondria possess efficient H_2O_2 scavenging systems and mechanisms to prevent free radical formation.

As a by-product of respiration, most mitochondrial ROS are generated in electron transport. This ROS production is associated with the metabolic conditions of the cell and generally occurs at higher levels compared to the levels ROS produced in the cytosol.

Iron is an important component of free radical biological oxidation, one of the mechanisms is through the *Fenton reaction* and the *Haber-Weiss reaction* catalyzed by iron (reactions 10 and 11, chapter 1). An alternative pathway is iron-mediated lipid peroxidation (Chapter 1) and not necessarily starting with the ·OH formed from the Fenton or Haber-Weiss reaction. One of the key factors in cell damage is iron-mediated lipid peroxidation.

Reactive oxygen species - ROS (O^-_2, ·OH, NO·, RO·, ROO·) are produced by cells from aerobic metabolism. They are involved in cellular redox metabolism processes, intracellular signaling, synthesis of biological compounds, phagocytosis, cell proliferation and ATP synthesis, among others.

Superoxide ion is also produced during reperfusion of oxygenated blood into tissues that are briefly anoxygenic. Organs that remain in anoxygenic conditions for some time may present damage from severe to serious that can be mitigated or prevented if SOD or other protective substances are added to the blood during reperfusion. One of the possible mechanisms of superoxide ion formation is when xanthine oxidase (XO) is produced from xanthine dehydrogenase in a reaction catalyzed by an activated proteinase during anoxia conditions. Xanthine oxidase reduces the O_2 to O^-_2 and H_2O_2 [5].

The redox pairs that supply the reducing capacity with reduction potentials associated to the system are the same that provide cells and tissues redox environment. Thus, in mitochondria, the passage of electrons from high energy bonds to O_2, by deviation, leads to the superoxide and peroxide production, affecting the signaling pathways in biological systems.

The cells vary in the rate of oxygen consumption, in the levels of antioxidants and in the redox enzymes through which the cellular redox environment is maintained. Further quantitative information on the redox enzymes and metabolic species involved is necessary for a complete understanding of cell and tissue redox biology.

CONCLUSION

Oxygen is incorporated into the metabolism of living organisms through oxidation- reduction reactions by enzymes. An analysis was made of the changes in free energy, expressed as redox potential (E'0), of the enzymes involved in biological oxidation-reduction processes, which exert their activity through the transfer of electrons or protons. Some reactive oxygen species - ROS are produced as a product of normal oxygen metabolism in the mitochondrial electron transport chain - ETC.

REFERENCES

[1]   Murray, R.K.; Granner, D.K.; Mayes, P.A.; Rodwell, V.W. *Harper's Ilustrated Biochemistry,* 26th ed; Lange Medical Books/ McGraw-Hill, **2003**.

[2]   Wikström, M.; Krab, K.; Sharma, V. Oxygen activation and energy conservation by cytochrome *c* oxidase. *Chem. Rev.,* **2018**, *118*(5), 2469-2490.
       [http://dx.doi.org/10.1021/acs.chemrev.7b00664] [PMID: 29350917]

[3]   Alfonso-Prieto, M.; Biarnés, X.; Vidossich, P.; Rovira, C. The molecular mechanism of the catalase reaction. *J. Am. Chem. Soc.,* **2009**, *131*(33), 11751-11761.
       [http://dx.doi.org/10.1021/ja9018572] [PMID: 19653683]

[4]   Denisov, I.G.; Makris, T.M.; Sligar, S.G.; Schlichting, I. Structure and chemistry of cytochrome P450. *Chem. Rev.,* **2005**, *105*(6), 2253-2277.
[http://dx.doi.org/10.1021/cr0307143] [PMID: 15941214]

[5]   Kowaltowski, A.J.; de Souza-Pinto, N.C.; Castilho, R.F.; Vercesi, A.E. Mitochondria and reactive oxygen species. *Free Radic. Biol. Med.,* **2009**, *47*(4), 333-343.
[http://dx.doi.org/10.1016/j.freeradbiomed.2009.05.004] [PMID: 19427899]

CHAPTER 3

Reactive Oxygen Species Sources

Abstract: Reactive oxygen species -ROS are produced by the oxidation-reduction processes of some enzymes as a product of the energy metabolism of cells and organelles and by exogenous sources. The mechanisms of oxidation-reduction and the enzymatic kinetics of some ROS source enzymes are presented here, as well as the mechanisms related to ROS production in organelles and cells and exogenous sources such as Fe, Cd, Hg, Ni, Zn, among others and xenobiotics such as paraquat-PQ.

Keywords: *Cytochrome C*, Electron transfer, Enzymatic kinetics, Enzymatic oxidation, Galactose Oxidase, *Haber – Weiss* reactions, Heavy metal contamination, Microsomes, Paraquat, Reaction centre, Xanthine oxidase, Xenobiotics.

1. ENDOGENOUS SOURCES

Oxygen-free radicals can be produced in cells by various processes and reactions: Radiations on photosensitizers such as retinal, riboflavin, chlorophyll or bilirubin, redox reactions with transition metals or redox reactions catalyzed by enzymes.

1.1. Enzymes

Some enzymes generate oxygen free radicals during their catalytic cycle; therefore, regulating the activity of these enzymes can control the oxygen free radical concentration. Monoamine oxidase deaminates dopamine and forms H_2O_2 in neurons. Aldehyde oxidase oxidizes aldehydes in the liver and releases O^{-}_2. Cyclooxygenase and lipoxygenase are enzymes of the biosynthetic pathway of prostaglandins, thromboxanes and leukotrienes, also release oxygen-free radicals. These radicals can inactivate the enzymes that originate them and thus regulate the route in which they participate. On the other hand, it has been demonstrated that cyclooxygenase is also capable of metabolizing certain xenobiotics to more toxic species, which may react with molecular oxygen to give rise to new oxygen reactive species.

In addition to the above, xanthine oxidase is an important source of reactive oxygen species -ROS. This enzyme, under normal conditions, presents dehydrogenase activity and oxidizes xanthine to uric acid utilizing NAD^+. When energy charge decreases for example, because of ischemia, the enzyme functions as oxidase, utilizes molecular oxygen to oxidize its substrate and produces superoxide radical and hydrogen peroxide. Calcium channel dysfunction during an ischemic period releases calcium ions from their reservoirs and these activate proteases catalyze the conversion of xanthine dehydrogenase to xanthine oxidase. Xanthine oxidase does not act as such until reoxygenation (Fig. **1**).

Another important source of oxygen free radical is NADPH oxidase. When foreign particles invade the body, the inflammatory response is triggered. During the process, macrophages, and neutrophils, activated by contact with the foreign substance, increase their O_2 consumption, which is transformed into O^-_2, which is then converted into hydroxyl radical and hydrogen peroxide.

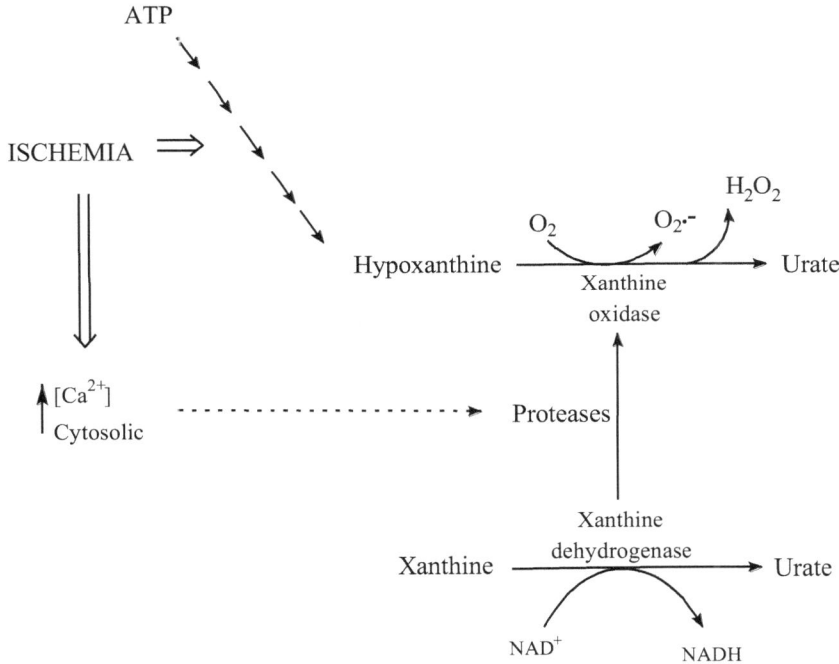

Fig. (1). Reactive oxygen species production during the ischemia process.

This respiratory flame is due to the enzyme complex located on the outer side of the plasma membrane and known as NADPH oxidase. NADPH oxidase contains adenindinucleotide flavine (FAD) and a cytochrome b_5 type with a sufficiently low redox potential to reduce O_2 to O^-_2, with NADPH electrons. In this way, foreign particles are exposed to oxygen free radical toxicity in the phagocytic vacuole [1]. The lysosomal myeloperoxidase released in the vacuole forms hypochlorous acid in the presence of halides and H_2O_2 (Fig. **2**). This acid is very reactive and can oxidize different biological molecules; in also, it can react with O_2^- to produce ·OH, or with H_2O_2 to produce singlet oxygen.

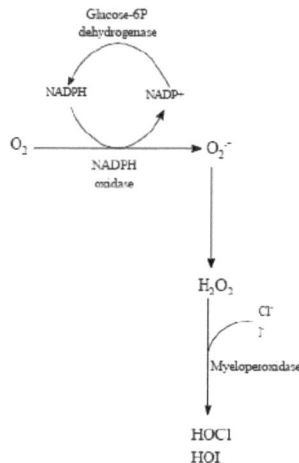

Fig. (2). Reactive oxygen species production during respiratory flame.

1.1.1. Cytochrome c – Enzymatic Oxidation

Cytochrome *c* (Cyt *c*) is found in the mitochondrial inner membrane, specifically on its outer surface. Cyt *c* is a 12kDa globular protein with a *heme* group, protoporphyrins IX and Fe^{3+}, covalently linked to two cysteine residues Cys-14 and Cys-17, coordinated to His-18 and Met-80. In mammals, there are 104 amino acid residues. Cyt *c* reacts with the hydroperoxides of α-linolic acid, leading to its decomposition by homolytic cleavage and producing singlet oxygen.

Cytochrome *c* and phospholipids are major targets of ·OH, hydroxyl and O_2^- superoxide radicals during oxidative stress. Protoporphyrin IX and amino acid chain residues of Cyt *c* that are hydroxylated or carboxylated are oxidized during hydroxyl- and superoxide radical-mediated oxidation of Cyt *c*. Among these amino acid residues is Met-80, which is oxidized to methionine sulfide, becoming the main ROS target. The derangement from the coordinated Met-80/iron *haeminic* bond leads Cyt *c* from being an electron transporter to being a peroxidase.

The free radicals formed in enzymatic oxidation have high reactivity and act as reductants or oxidants. These reaction rates can be measured by electron spin resonance -ESR, a spectrophotometer and flow equipment. Thus, Ohnish *et al.*,

1969 [2], by these techniques measured the reaction constants between cytochrome b_5 and 2-methyl-1,4-naphosemiquinone -MKH and the rate of reduction of cytochrome c by p-benzosemiquinone and monodehydroascorbate. They also determined by ESR spectroscopy that free radicals derived from chlorpromazine and p-cresol are one-electron oxidants.

During the benzohydroquinone peroxidative oxidation reaction, the cytochrome c is reduced. In the presence of considerable amounts of cytochrome c and when the supply of p-benzoquinone is low, between the rates of benzohydroquinone oxidation and cytochrome c reduction there is a stoichiometric balance.

Therefore, occurs the following reaction:

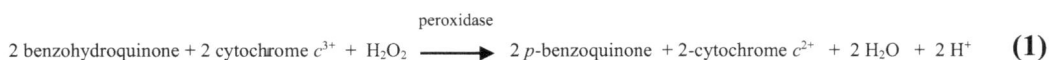

$$2\ \text{benzohydroquinone} + 2\ \text{cytochrome}\ c^{3+} + H_2O_2 \xrightarrow{\text{peroxidase}} 2\ p\text{-benzoquinone} + 2\text{-cytochrome}\ c^{2+} + 2\ H_2O + 2\ H^+ \quad \textbf{(1)}$$

The one-electron oxidation of H_2A donor molecules to form $HA^{\cdot}$ monodehydro molecules is catalyzed by peroxidase.

$$2H_2A + H_2O_2 \xrightarrow{\text{peroxidase}} 2\ HA^{\cdot} + 2\ H_2O \qquad \textbf{(2)}$$

$$2HA^{\cdot} \longrightarrow H_2A + A \qquad \textbf{(2')}$$

Through the reaction with cytochrome c, the $HA^{\cdot}$ formed during the peroxidase reaction (reaction 2) decreases , as demonstrated by the stoichiometry of reaction 1. Therefore, cytochrome c reduction constant is given by *equation* 1, considering that at pH 6.5 cytochrome c reduction by benzohydroquinone is slow.

$$\frac{d\ \text{cytochrome}\ c^{2+}}{dt} = k_r\ (p\text{-benzosemiquinone})\ (\text{cytochrome}\ c^{3+}) \qquad \textit{Equation 1}$$

The value obtained by indirect methods in non-enzymatic systems is consistent with the value obtained for k_r, which is $2.5 \times 10^6\ M^{-1}sec^{-1}$. The one-electron oxidation of ascorbate and reductase is catalyzed by ascorbate oxidase enzyme.

Cytochrome c may be reduced by the previously formed monohydroascorbate, which is not active as p-benzosemiquinone or MKH (2-methyl-1-4-naphthosemiquinone). For this cytochrome c reduction reaction, the constant is $4 \times 10^4\ M^{-1}sec^{-1}$. The monodehydro forms of NADH and indolacetate are active intermediates considering that these molecules are substrates for the peroxidase oxidase reaction.

There is minimal reduction of iron by NADH, suggesting that it is carried out by the following mechanisms:

$$2\,NADH \;+\; H_2O_2 \xrightarrow{\text{peroxidase}} 2\,NAD\cdot \;+\; 2H_2O \tag{3}$$

$$2\,NAD\cdot \;+\; 2Fe^{3+} \longrightarrow 2NAD^+ \;+\; 2Fe^{2+} \tag{4}$$

Metmyoglobins, cytochromes b_5 and c, peroxidases, and O_2 can also be reduced by NAD· generated during peroxidative oxidation.

The mechanisms in Scheme **1** are supported by kinetics and stoichiometry the reactions in which enzymes catalyze the transfer of electrons from donors and the main pathway for radical disappearance is dismutation or dimerization.

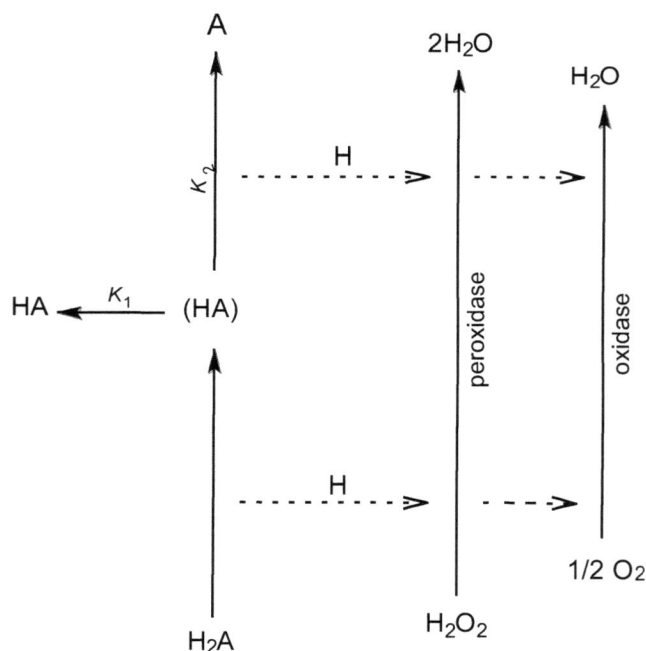

Scheme (1). Two electron (K = 0) and one electron (K = 2) transfer at the surface of oxidative enzymes.

Value of κ is determined in the following *equation*

$$\kappa = \frac{k_1}{\frac{1}{2} k_1 + k_2} \qquad \textbf{\textit{(Equation 2)}}$$

When κ is near 2, the mechanism of one-electron passage may be involved in reactions such as those of ascorbate oxidase and peroxidase, whereas when κ is near zero, the mechanism involved is two-electron passage as in catechol oxidation by tyrosinase.

By means of an ESR spectrophotometer, the second-order reaction constant k_r of *equation* 1 may be measured by the reaction rate of the cytochrome and by monodehydroxide concentration, which may be measured at a concentration of 0.1 to 1 µM. This is possible when monodehydro molecules specifically reduce cytochromes. The relationship between the second order reaction constant k_r and the half-life of the reduced cytochrome $(t_{1/2})$ is presented in *equation* 3.

$$t_{1/2} = \frac{0.693}{k_r(HA\cdot)} \qquad \textbf{\textit{(Equation 3)}}$$

The reaction constant from this *equation* is $10^9 M^{-1} sec^{-1}$ which corresponds to the upper limit measured by this method.

The exothermicity and selectivity of the reactions are presented in Scheme **2**. Since they are exothermic, the electron transfer reactions from monodehydro to cytochromes and from donor molecules to enzymes require little activation energy. This scheme illustrates that a strong one-electron reductant is produced by enzyme systems during the selective oxidation of one-electron donor molecules. In biochemical reactions, selectivity in the reaction is characteristic, which would explain the relatively easy cytochrome reduction under the right conditions and why monodehydro molecules may act as both oxidizers and reductants. In both cytochrome b_5 oxidation and NADH oxidation catalyzed in microsomes, it is possible that monodehydroascorbate is involved as an oxidant. In the scheme, the valence of the catalytic site is represented by n. The level of the redox potential is represented by the ordered position and the electron flux upstream is given as the exothermicity of the reaction.

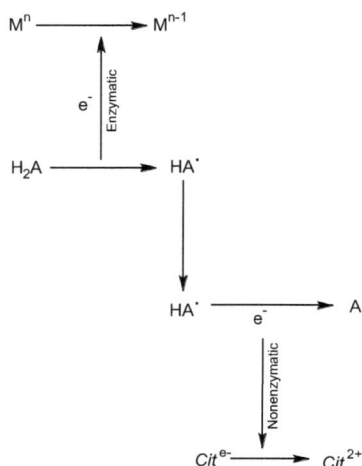

Scheme (2). Cytochrome reduction catalyzed by specific electron oxidants.

Strong one-electron oxidation by the monodehydro is generated when oxidogenic substrates are utilized as electron donors in the peroxidase system. The role of oxidogenic substrates in electron transfer is presented in Scheme **3**. HX indicates the activity of an oxidogenic substrate. Oxidogenic substrates such as monohydroxybenzene derivatives stimulate the removal of an electron in slow peroxidase substrates.

Ascorbate and NADH are oxidized by the monodehydro-*p*-cresol chlorpromazine. However, glutathione is only oxidized by the free radical chlorpromazine.

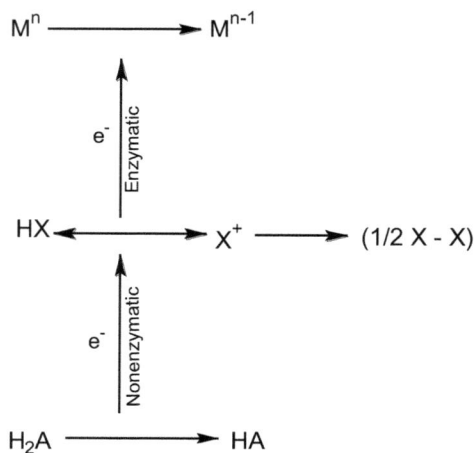

Scheme (3). One-electron HX-mediated transfer from H_2A to the enzyme.

The oxidative activity of hydroxyl and superoxide radicals obtained radiolytically in the presence or absence of Cyt c, on glycerophosphocholine liposomes -PLCP was evaluated by Sidahmed-Adrar *et al*, in 2010 [3]. They demonstrated that there is a protective effect of Cyt c to PLPC and that this is given by reactions of ·OH radicals towards conjugated dienes and hydroperoxides $PCOOH^T$. Although the reaction constant of glycerophosphocholine-PLPC is unknown, the reaction constant of Cyt c + ·OH was found to be 1.4×10^{10} $mol^{-1}s^{-1}$. On the other hand, hydroxyl radical action on PLCP may be effective when PLCP consumption is slow with Cyt c.

The radical reaction constant with phosphocholine is 6.4×10^8 $mol^{-1}s^{-1}$ and with linoleate is 9.0×10^9 $mol^{-1}s^{-1}$. In glycerophosphocholine liposomes -PLCP, the molecules are tightly packed, the linoleate residues are not very accessible and the polar moiety such as phosphocholine is exposed to water, so it may be subject to the action of ·OH radicals. Thus, oxidative fragmentation of PLCP occurs even in the presence of low concentrations of ·OH hydroxyl radicals, because phosphocholine is sensitive to the action of radicals. During liposome disintegration, ·OH and linoleate reactions are generated, producing conjugated dienes and $PCOOH^T$, because linoleate residues become accessible to hydroxyl radical action. Thus, hydroperoxides are converted to hydroxides very rapidly when PLCP oxidation occurs in the presence of Cyt c.

Fig. (**3**) represents the five redox reactions involved in the reaction of cytochrome c and $PCOOH^T$ to PCOH. In reaction 5, through electron reduction, hydroperoxides are converted to alkoxyl radicals PCO· producing as an intermediate an oxoferryl cation FeO^{2+} or Fe (IV) = O. Also, through two-electron reduction, they are converted to hydroxyl involving the oxidation of iron and its oxoperferyl state FeO^{3+} or Fe (V) = O (reaction 6). By hydrogen extraction, the alkoxyl radicals produced in reaction 5 react with PLCP and thus amplify the PLPC degradation. The above mechanism may only occur during incubation with Cyt c, for a small percentage of the total PCOH.

During the peroxidase activity of Cyt c, it acts as a catalyst, thus transforming six molecules of $PCOOH^T$, in a reaction involving the reduction of *heme*-FeO^{2+} and/or *heme*-FeO^{3+} to *heme*-Fe^{3+}. Hydroperoxides are the oxidizable substrates for the reduction of the oxoferryl cation (reaction 7) as present in myoglobin. The peroxyl radical from PCOOH can be converted again to another PCOOH with the regeneration of ferric iron. In turn, the superoxide radicals generated by radiolysis of water reduce the ferric iron from Cyt c to ferrous iron (reaction 8), in a reduction that does not affect the peroxidase action of Cyt c to $PCOOH^T$. For its part ferrous iron participates in alkoxyl radical formation, mediating the peroxidase activity of PCOOH transformation (reactions 9 and 10) or with different oxidation states, may participate in PCOH formation (reactions 5 and 6). Therefore, by one- or two-electron reactions involving ferrous or ferric irons, the peroxidase activity of Cyt c to PCOOH^T could be regulated.

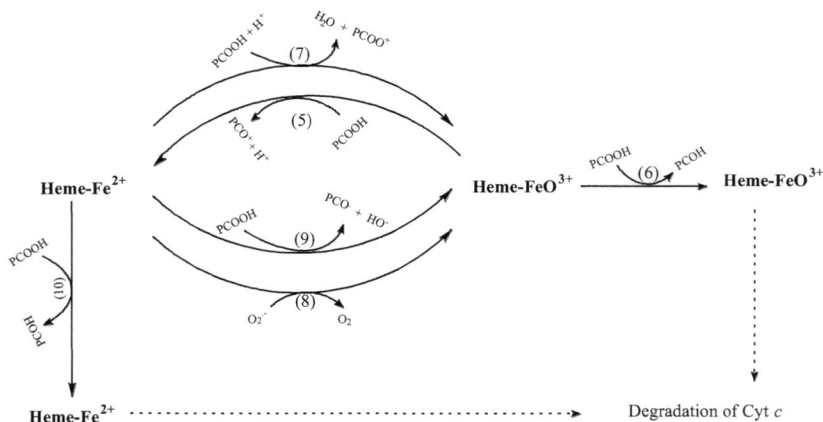

Fig. (3). Scheme of conversion mechanisms from PCOOH to PCOH activated by Cyt c. Reactions are explained in the text.

Strong oxidations are generated by the iron intermediate species ferril (+4) and perferril (+5), which lead to the degradation of Cyt c. In addition, these species can also degrade tryptophan and tyrosine residues, considering that Cyt c contains one tryptophan and four tyrosines. Similarly, the oxoferryl cation could oxidize protoporphyrin IX of Cyt c.

1.1.2. Xanthine Oxidase (XO) -Reaction Center

Among the molybdenum hydrolase family is xanthine oxidase (XO). Mammalian XO is a 290 kDa homodimer with one flavin, one molybdenum, and two iron-sulfur [Fe_2S_2] centers per subunit. Although its physiological function is the oxidation of purines, it can also catalyze the oxidation of various aldehydes, the hydroxylation of purines at various ring positions such as 2, 6 and 8 and the oxidation of other substrates.

XO as a source of oxygen for substrate hydroxylation utilizes water instead of oxygen and generates reducing equivalents during catalysis, unlike monooxygenase activity. At the molybdenum site that is reduced from Mo^{VI} to Mo^{IV}, substrate oxidation takes place. An electron transfer from molybdenum to [Fe_2S_2] groups completes the catalytic cycle. In the [Fe_2S_2] groups, the electrons are donated to an acceptor such as O_2. Molybdenum is covalently bound to the dithiolene moiety from the organic cofactor molybdopterin.

In the structure of the active site of XO (Fig. **4**) the molybdenum atom is bound by two long thiolate ligands (Mo-S), a short terminal sulfur unit (Mo=S), and a short oxygen (Mo=O).

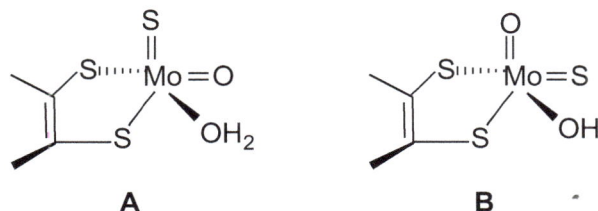

Fig. (4). Xanthine oxidase active site. Crystallographic, A; Spectroscopic, B.

At the active site of xanthine oxidase, the catalytic oxygen donor is Mo-OH instead of Mo-OH$_2$ as shown in Fig. (**5**). This was demonstrated by X-ray crystallography studies realized by Donaan *et al.*, 2005 [4]. In the substrate oxidation that occurs during catalysis, at the catalytic site of XO, Mo^{VI} is reduced to Mo^{IV} followed by reoxidation involving a transient Mo^V intermediate. This reoxidation is enhanced by proton transfer and electron coupled reactions. The initial reductive events are catalyzed with Glu 1261 acting at the base of the active site, considering pH dependence of the reaction. An important nucleophile initiates the catalytic sequence when acting on the C-8 position of the substrate. This nucleophile is formed when deprotonation occurs at the base from Mo-OH to Mo-O$^-$ instead of Mo-OH$_2$ to Mo-OH.

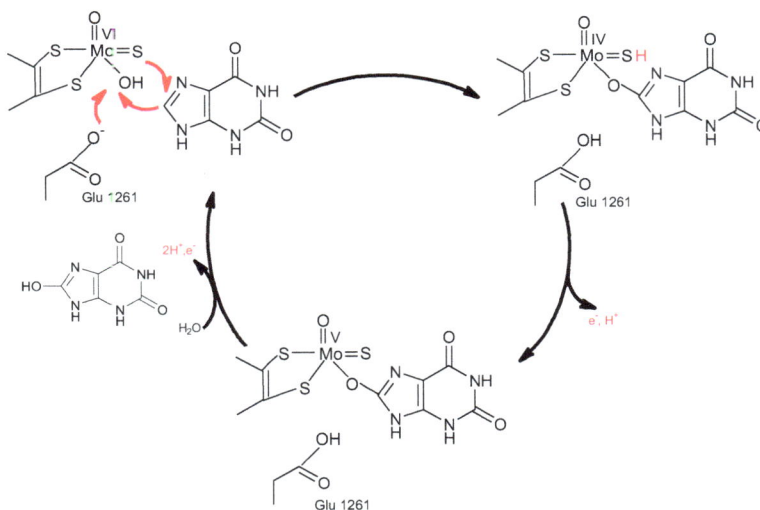

Fig. (5). The catalytic mechanism for Xanthine oxidase.

In proposed schemes for the Mo center function in molybdenum hydroxylase catalysis, the Mo simultaneously functions as an electron and proton receptor. Similarly, it has been demonstrated that hydrogen is transferred directly from the C-8 of substrates to the Mo center of xanthine oxidase.

Under a series of experimental conditions, Olson *et al.*, 1974 [5] determined that internal electron transfer is not rate-limiting for catalysis among the various reducible centers present in xanthine oxidase. An internally balanced electron pool connects these centers. From the balance between the potential reductions of the individual centers and the number of electrons in the enzyme, the electronic distribution between the various centers of the enzyme could be determined.

The reduction potentials are pH dependent. The alteration of the reduction potentials is caused by the change in the direction of electron transport and by the decrease or detention of the electron flow. These changes are occasioned by the difference in proton or chemical concentrations that interact with the resting states of the electron transport chain.

In studies realized by Barber and Siegel, 1982 [6], the reduction potentials as a function of pH, for the xanthine oxidase Mo center and for the FAD and Fe/S centers have been examined.

Molibdene: Mo reduction may be accompanied by Mo center protonation. In studies carried out the potential *vs.* pH curves demonstrate that at a wide pH range, the Mo^{VI} to Mo^V and Mo^V to Mo^{IV} conversion reactions are accompanied by the addition of protons. Under pH ranges between 6.0 and 10.9 the Mo^V centers remain protonated for both disulfide xanthine oxidase and native xanthine oxidase.

The reduced Mo center of the xanthine oxidase disulfide enzyme has a higher affinity for protons than the native enzyme Mo^{IV}, so that at pH values close to 7.0 the Mo^V reduction involves the conduction of one proton in the native enzyme and two protons in the disulfide enzyme. In the pH range from 6.0 to 10.9 there is independence of the Mo^{VI}/Mo^V potentials of the terminal ligand of Mo bound to the enzyme (oxygen *vs.* sulfur).

The reduction potentials of Mo^{VI} to Mo^V and Mo^V to Mo^{IV} are not equivalent in molybdenum enzymes. Thus, at alkaline pH, the potentials E_{m1} and E_{m2} in the native oxidase are different. Thus, at pH 10.9 a very small EPR signal of Mo^V may be observed compared to those observed at pH 7.7.

FAD: Xanthine oxidase is characterized because it has NAD^+ as an acceptor with a low substitution number, and a high substitution number for the oxidation of O_2

as an electron acceptor. This disables it to stabilize the semiquinone state of FAD. On the other hand, having NAD^+ with a high substitution number and O_2 with a low substitution number is characteristic of the dehydrogenase enzyme.

At pH 7.0 about 20-30% of the FAD enzyme can be present in the semiquinone state. It has an absorbance of 580-660 nm as detected by EPR. At pH range 8-9, the potentials E_{m1} and E_{m2} are highly divergent and small amounts of semiquinone have been detected there by EPR. There is no relationship between the amount of semiquinone FAD and the amounts of absolute or relative activity with O_2 vs. NAD^+ when the concentration of xanthine oxidase with O_2 and NAD^+ act as a function of pH. Therefore, the main difference between the specificity of oxidase and dehydrogenase is that there is no ability to stabilize FAD in the one-electron oxidation state. In xanthine oxidase, interconvertibility between the anionic semiquinone and the neutral flavin can be demonstrated.

Iron-sulfur Centers: At very high or very low pH values the E_m potentials at the Fe-S centers are pH independent. However, in some cases even the E_m potentials are pH dependent although the maximum slope of E_m with respect to pH was considerably less than -60 mV per pH unit. This occurs when an Fe-S center is oxidized or reduced or when there is a difference in the affinity of a single proton, since enzymes with reduced Fe-S centers are more strongly binding protons than enzymes with oxidized Fe-S centers. The variation of the E_m potential with respect to pH is relatively slight when the pk for the dissociation of this proton is increased by one unit on addition of the electron to the Fe/S_{II}. In contrast, the variation of E_m with respect to pH in the Fe/S_I center reduction is much larger when the pk for the dissociation of the proton increases by 1.5 units.

The balance in the reduction processes at the centers changes according to pH due to variations in the reduction potentials of flavin and Fe/S. The Fe-S center reduction occurs before the FAD reduction at pH 6.1 while the Fe/S_I or FAD reduction can take place after the Fe/S_{II} center reduction at pH 8.9.

The mechanisms by which xanthine oxidase catalyzes at the same time in *p*-benzoquinone one electron and two electron reduction was analyzed by Nakamura and Yamazaki, 1969 [7]. They determined that cytochrome reduction is accomplished by oxygen-mediated electron transfer and *p*-benzoquinone. Similarly, perhydroxyl radicals and semiquinone are also direct reductants of cytochromes in the xanthine oxidase system.

From some aldehydes and purines the cytochrome c reduction, $Fe(CN)_6^{3-}$, O_2, NO_3^-, dyes and some quinones is catalyzed by xanthine oxidase. Electron transporters from enzymes to cytochrome c can be carried out by compounds such as 2-methyl-1,4-naphthoquinone-MK and from O_2 presence depend on the cytochrome c reduction by xanthine oxidase. The perhydroxyl anion O_2^- is the reductant of cytochrome c and is produced when the enzymes dihydroortic acid dehydrogenase, aldehyde oxidase and xanthine oxidase catalyze the reduction of O_2. Thus, the perhidoxyl radical is involved in the oxygen-mediated transfer of electrons from xanthine oxidase to cytochrome c. Furthermore, cytochrome c reduction is carried out by xanthine oxidase in the presence of p-benzoquinone under anaerobic conditions.

ESR is employed to measure the disappearance of p-benzoquinone under experimental conditions. Thus, the steady-state p-benzosemiquinone concentration depends on the rate of p-benzoquinone reduction (v) as described in equation 4:

$$[p\text{-benzosemiquinone}]_s = (K \cdot V / 2k_d)^{1/2} \qquad \textit{(Equation 4)}$$

Where κ is the specific constant for the enzymatic reaction and k_d is the dismutation constant of p-benzosemiquinone. On the concentration measure, the rate of p-benzosemiquinone reduction (v) is constant. Cytochrome c is at the appropriate concentration to trap all the p-benzosemiquinone formed, it could be demonstrated if the initial rate of reduction of cytochrome c mediated by p-benzoquinone will give rise to κ. Considering the above, then the initial rate of cytochrome c reduction and the rate of p-benzosemiquinone formation, could be equal $K \cdot V$.

For κ values lower than 2, benzohydroquinone acts as a strong reductant of cytochrome c, so that from the κ value and the p-benzoquinone concentration depends on the net amount reduced from cytochrome c. On the contrary, at $\kappa = 2$ benzoquinone acts as an electron transporter from xanthine oxidase to cytochrome c.

The net amount of reduced cytochrome c (S) in the presence of an adequate concentration of cytochrome c can be expressed by the following equation 5:

$$S = \frac{av}{1 - r} \qquad \textit{(Equation 5)}$$

Where r is a common radius of the geometric series corresponding to the ratio between k_1 and $(k_1 + k_2)$ and a corresponds to the concentration of p-benzoquinone added. In Scheme **4** the constants k_1 and k_2 are defined and the value of κ is given by:

$$K = \frac{k_1}{\frac{1}{2}\,k_1 + k_2} \qquad\qquad \textbf{\textit{(Equation 6)}}$$

Considering that r is variable and may decrease during the reaction, the к values obtained from the net amount of reduced cytochrome c can only be qualitative.

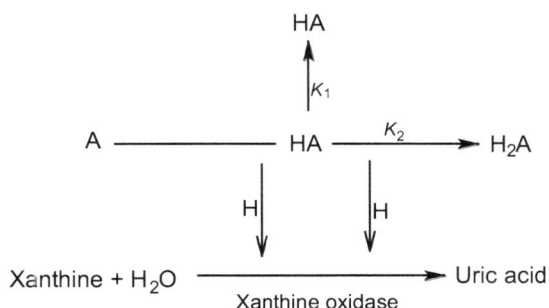

Scheme (4). Reaction mechanism between xanthine oxidase and the acceptor of two electrons. The k_1 to k_2 ratio is the electron acceptor concentration dependent A.

If к = 2 as in the reaction of NADPH and cytochrome c reductase, the molar ratio could be 2 for compound III of lactoperoxidase formed by the addition of xanthine. Similarly, if the reaction of the perhydroxyl radical with peroxidase is unfavorable at high pH, a decrease in the efficiency of compound III formation could occur at these pH levels. When the method employed is a quantitative perhydroxyl radical assay, this pH limitation represents a difficulty.

From the point of view of redox activity, the radical thus formed acts like a semiquinone. At pH 7.0 the redox potential is 0.27 V for the O_2/H_2O_2 pair. At pH 7.0 for the pairs:

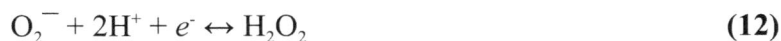

$$O_2 + e^- \leftrightarrow O_2^- \qquad\qquad \textbf{(11)}$$
$$O_2^- + 2H^+ + e^- \leftrightarrow H_2O_2 \qquad\qquad \textbf{(12)}$$

The redox potential for reaction 11 is between 0.3 and 0, 5V and for reaction 12 the redox potential is between 0.84 and 1.04 V. The reaction 11 redox potential does not depend on pH since the pK of the perhydroxyl radical is 4.5. Similarly, according to reaction 11, the perhydroxyl radical acts as a reductant because in the presence of a suitable electron acceptor the redox potential is less than -0.3. The perhydroxyl radical may reduce cytochrome *c* or react with peroxidase to form compound III, when formed on the surface of the enzyme and is therefore enzyme-free.

Scheme **5** presents the electron transfer pathways from xanthine oxidase to cytochromes *via* molecular oxygen, *p*-benzoquinone, and MK.

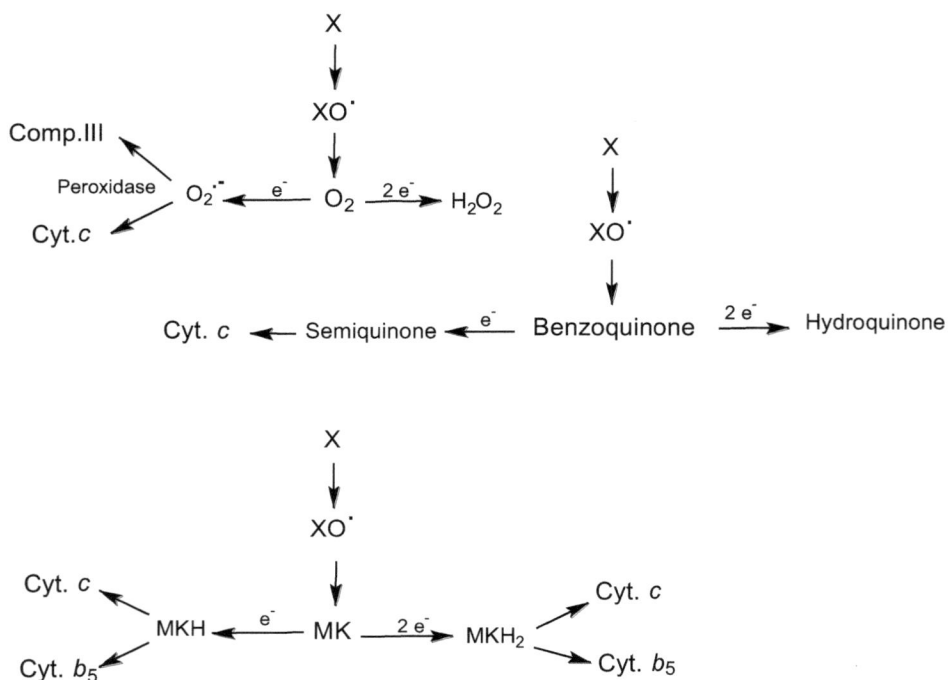

Scheme (5). Electron transport pathways from Xanthine oxidase to cytochromes.

An electron transporting agent attached to external acceptors such as O_2, could be iron. *p*-benzoquinone, acts in the same way as molecular oxygen [7].

Heuvelen, 1976 [8] based on the properties of xanthine oxidase as relative reduction potentials, electron transfer and separation of the enzyme's resting sites, proposed a tunneling model to explain the electron transfer mechanism, considering the mechanisms that allow the stability of the electrons in the different sites of the enzyme.

The electron resting sites per active center in the xanthine oxidase system may accept electrons from three substrates, *i.e.*, six electrons: one electron each for the Fe/S sites (Mo^{5+}, $FADH^{\cdot}$, $Fe/S_I^{\cdot}$ and $Fe/S_{II}^{\cdot}$) and two electrons each for the molybdenum and flavin sites. The rest sites with an extra electron are paramagnetic.

The electron energy has been calculated by comparing the distribution probabilities for two electrons, one-50%, four-50% and six electrons. It has been found that, among the different active sites, the electrons are distributed unequally. Thus, with a measure of two electrons per active site in enzymes, these electrons are rearranged in such a way that some sites have one, while others have three, etc. Depending on the number of pathways for each situation and the relative energies each active site moiety chooses zero, one, two, three or eight electrons.

The relative reduction potentials are at 160 mV at each of the different centers. With the energies presented in Table **1** [8], schemes can be developed for the transfer of two electrons from Mo^{4+} to $FADH_2$, in the jump of an electron that involves only a 100 mV change of potential per jump, since some jumps present few energy changes.

Table 1. Occupations of electrons in different resting places.

pH	No. of electrons per active site	Probabilities of observed and (calculated*) electron occupancy				
		Mo^{5+}	FADH*	Fe/S_I*	Fe/S_{II}*	$(Mo^{4+} + FADH_2)$**
6.3	1	0.06(0.05)	0.02(0.04)	0.31(0.32)	0.61(0.59)	(0.00)
6.3	2	0.11(0.09)	0.03(0.04)	0.50(0.37)	0.65(0.41)	0.36(0.04 + 0.49)
6.3	(4+6)/2	0.14(0.15)	0.04(0.01)	0.72(0.86)	0.86(0.92)	1.62(0.55 + 0.98)
10.1	1	0.11(0.10)	0.03(0.04)	0.24(0.33)	0.61(0.59)	(0.00)
10.1	2	0.14(0.15)	0.04(0.07)	0.40(0.33)	0.55(0.41)	0.44(0.03 + 0.49)
10.1	(4+6)/2	0.12(0.09)	0.02(0.01)	0.90(0.90)	0.90(0.92)	1.53(0.58 + 0.96)

* Calculations based on energies.
** The number of electrons in species $Mo^{4+} + FADH_2$ is twice the number presented here.

In the xanthine oxidase system, the starting point of the electron transport chain $FADH_2$ presents an energetically favorable state such that, observing the lower energy states, electrons can find their way from Mo^{4+} to $FADH_2$.

It has been estimated that the time required for electrons transport between the different states of rest t_{et} can be 5 x 10^3 s.

1.1.3. Xanthine Oxidoreductase (XOR)

Xanthine oxidoreductase-XOR is a flavomolybdenoenzyme complex. In man, it is the terminal enzyme of purine catabolism, catalyzes xanthine to urate and hypoxanthine to xanthine. It is present in milk and some other tissues. XOR is a 300 kDa homodimer, each subunit containing one FAD site, two Fe_2S_2 sites, four *viz* redox centers and a molybdenum cofactor (Mo-Co). Mo-Co is an organic derivative of molybdopterin with a cyclized dithiolene side chain and a molybdenum atom (Fig. **6A**). In turn, Mo is pentacoordinated by two dithiolene sulfide atoms of molybdopterin, sulfur and two oxygen atoms (Fig. **6B**).

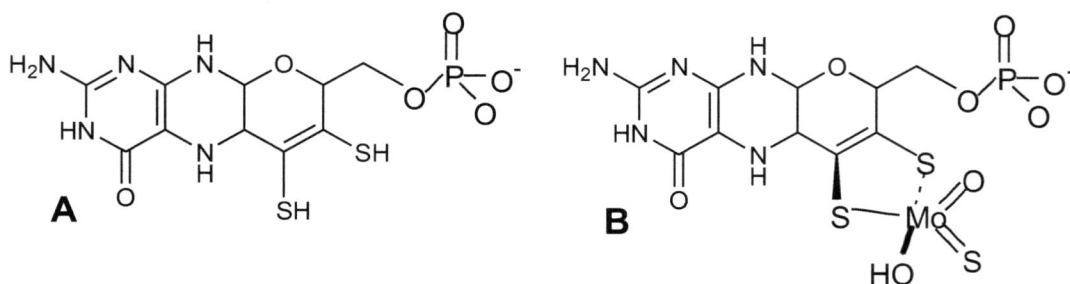

Fig. 6 (A). Molybdopterine. (**B**). The molybdenum cofactor.

In mammals, the enzyme exists in two interconvertible forms xanthine oxidase (XO; 1.1.3.22) and xanthine dehydrogenase (XDH; EC 1.1.1.204) found mainly *in vivo*. The interconversion of these enzymes can be performed XDH to XO by proteolysis in an irreversible manner or by sulfur reagents in a reversible manner. In contrast, in birds, only the XDH form is present. XO prefers molecular oxygen, unable to reduce NAD^+, whereas XDH preferentially reduces NAD^+. The ability of XOR to generate ROS consists in superoxide and hydrogen peroxide formation from the reduction of oxygen by one of the forms of the enzyme [9].

XOR also catalyzes *N*-heterocyclic substrates and aldehydes as does xanthine and hypoxanthine, in addition to acting as an NADH oxidase. In the Mo site, most substrates act, although only NADH donates its electrons to FAD (Fig. **7**). The electrons pass through the FAD center to NAD^+ or molecular oxygen in the redox balance of the centers. For oxidized substrates including methylene blue, ferricidin, 2,6-dichlorophenolindophenol and some quinones the specificity is low in a similar manner as for reduced substrates.

Hypoxanthine   Xanthine

Xanthine   Mo   Urate

Fe_2S_2

NAD^+ O_2^- H_2O_2

NADH

FAD

NADH O_2

NAD^+

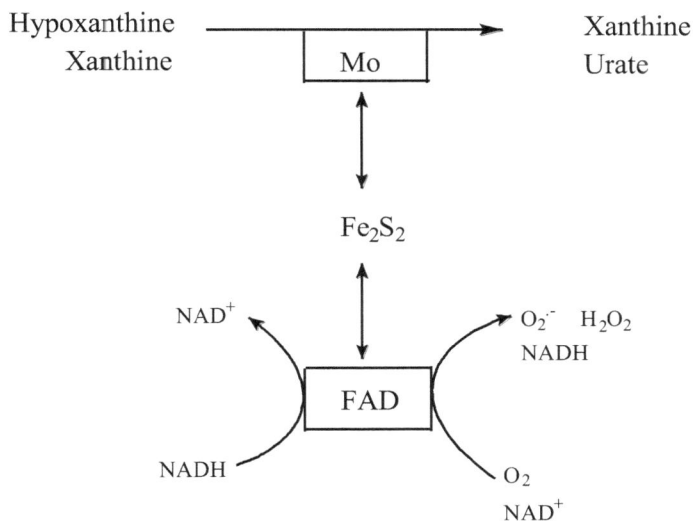

Fig. (7). Oxidation of NADH in the FAD site and xanthine and hypoxanthine in the Mo site.

1.1.4. Galactose Oxidase

A free radical metalloenzyme is the enzyme galactose oxidase, which contains a protein radical coordinated to a copper ion in the active site, forming a metal-radical complex. From the redox active side chain of the cross-linkage between the Tyr-Cys residues, the unusually stable radical protein is formed. Therefore, galactose oxidase belongs to a family of copper radical oxidase enzymes since it possesses an oxidation mechanism based on the copper-coupled catalytic motif of this free radical.

Galactose oxidase catalyzes the oxidation of the primary alcohol to an aldehyde through an O_2-coupled reaction to form hydrogen peroxide:

$$RCH_2OH + O_2 \rightarrow RCHO + H_2O_2 \tag{13}$$

The catalytic reaction could be equivalent to dihydrogen transfer between the two substrates because O_2 reduction and alcohol oxidation are processes involving two electrons. Organic redox cofactors such as flavins, quinones and nicotinamide are generally involved in biological hydrogen transfer. However, galactose oxidase contains as an active site a copper-coupled free radical complex but does not contain some of the redox cofactors mentioned above. Thus, the structure of the active site of galactose oxidase implies that the redox mechanism involves the two-electron reactivity of the metal-radical complex and the free radicals.

The active site of galactose oxidase contains a copper complex bound to four amino acid side chains: two histidines and two tyrosines. The tyrosyl-cysteine (Tyr-Cys) bond (Fig. **8**) is formed due to the tyrosine Tyr272 cross-linking the C_ε carbon of the phenolic side chain to the Sγ sulfur of Cys228.

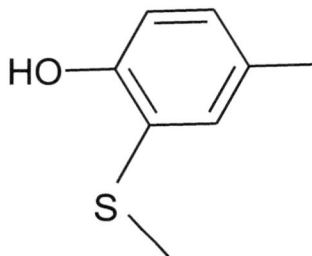

Fig. (8). Tyr-Cysteine group (Tyr-Cys) in galactose oxidase.

The protein structure and its reactivity are influenced by the thioether bond that exists between the two amino acid residues. Similarly, the effect of the disulfide bridge in such a protein thus contributes the cross-link to the rigidity of the active site. However, the thioether bond is formed irreversibly and is not sensitive to reductive cleavage, unlike the disulfide bridge. In the protein, the cross-link is formed spontaneously in the presence of both dioxygen and reduced copper $Cu^{1+.}$

The chemical reactivity of the side chain is altered, facilitating its oxidation by the presence of the thioether substituent in the cross-linked aromatic ring system Tyr-Cys. With respect to the corresponding unsubstituted phenol, the one-electron oxidation state of the thioether-substituted phenol is stable at 0.5 V, as demonstrated by photoelectron spectroscopy studies. The above result is based on the difference in the ionization thresholds of the molecules.

Quino cofactors during rotation form nucleophilic adducts with substrates and exhibit two-electron reactivity. In contrast, protein stabilization of free radicals in the active site of galactose oxidase is generated because the thioether substituent of the Tyr-Cys group is restricted to one-electron oxidation processes. Nucleophilic addition is not necessarily present in the Tyr-Cys ring system.

The direct interaction of Tyr-Cys with the metal in galactose oxidase distinguishes it from quino cofactors. The metal-radical complex in the active form of the oxidized enzyme (Fig. **9A**) is formed because Tyr-Cys binds to the active site of the metal ion, a redox active copper core. At room temperature and in the absence of reducing agents, this free radical-bound copper complex persists for weeks and is very stable. However, a catalytically inactive non-radical Cu^{2+} complex is formed (Fig. **9B**) by the presence of electron donors, reacting rapidly and generating a simple electron reduction. In addition, a reduced complex, which reacts with O_2 and constitutes a catalytic intermediate in the return cycle [10], is generated when the reduction converts the Cu^{2+} center to Cu^{1+} (Fig. **9C**).

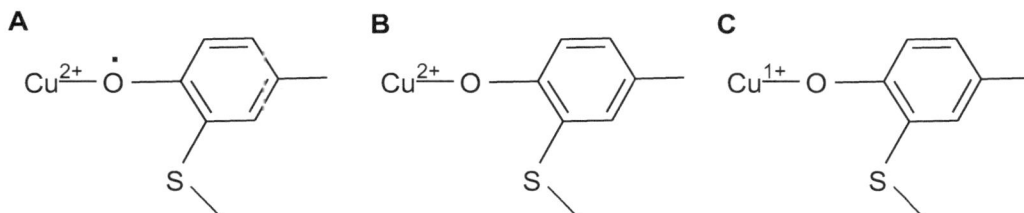

Fig. (9). Tyr-Cys group metal-radical complex in galactose oxidase. Explanation in text.

Like what occurs in the ligand exchange reaction during the return, acetate and amide ions displace water and coordinate directly with the metal center. In the water displacement the primary alcohols bind to the metal ions. Thus, acetate and amide interact more strongly with the metal cation than with neutral water. In the crystal structure of the anionic complex, the short bond distance to the exogenous ligand is due to the increased bond strength. A shortening of the Tyr 495 bond distance in the anionic adduct is due to an increase of the exogenous copper-ligand interaction that occurs *via* Tyr 495 interactions. Thus, in this complex, the coordination number of the metal ion decreases to 4. The plasticity of copper coordination chemistry causes reorganization of the inner sphere of the metal complex in response to changes in bond shortening.

The basicity of the phenolic side chain can be affected by the displacement of Tyr 495. Coordinated phenols are acidic and present high binding affinity to phenolate metals because of their altered pKs. Thus, a return to a value of pK= 10 is expected as metal interactions decrease.

The cuproxyl complex (Fig. **10D**) is the center of the catalytic complex in which the metal center is associated with a monoatomic oxygen free radical. It participates in the copper chemistry based on the Fenton reaction and oxidizes the protein at its side chains. This oxidant is "hot" E>1V. The free radical complex is stabilized by the electronic delocalization mechanism provided by the substitution of the phenyl ring (Fig. **10C**). It has been possible to synthesize and characterize related inorganic complexes, although no biological complexes with this structure are known. Additional stability to the free radical complex is provided by the thioether substitution of the phenoxyl ligand (Fig. **10B**) belonging to the Tyr-Cys site of galactose oxidase.

Considering the low redox potential $E_m = 0.45V$ of the galactose oxidase free radical, the metal-radical complex could be further stabilized by coordination through a second phenol group (Fig. **10A**). In the *bis*-phenol complex, stabilization is due to electronic delocalization that may occur by ligand-ligand charge transfer, inducing that on two identical ligands the unpaired electron is delocalized.

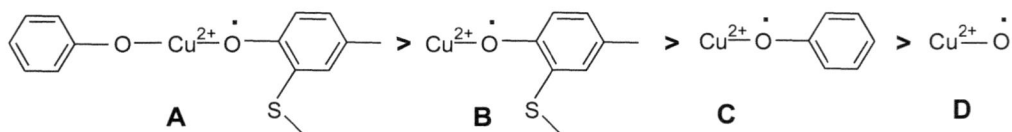

Fig. (10). Catalytic complex in galactose oxidase.

The delocalization pathway mediated by covalent interactions with the metal ion is illustrated by the superposition of two forms of limiting resonance (Fig. **11**). In galactose oxidase, for the electronic interactions indispensable for resonance stabilization, the geometry of the metal-radical complex is favorable.

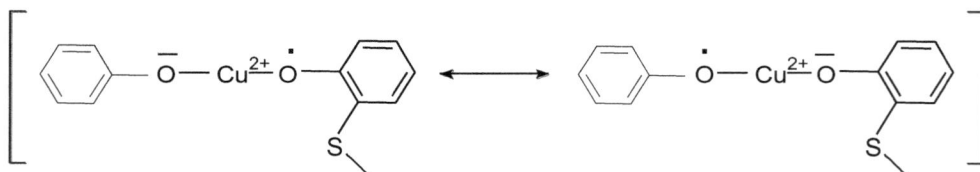

Fig. (11). Metal-radical complex geometry in galactose oxidase.

The possibility of reversible activation/inactivation by interconversion of the three oxidation states of the enzyme (Fig. **9 A-C**) difficult kinetic studies of galactose oxidase.

According to the ping-pong mechanism, reaction 13 can be written as separate reduction and reoxidation processes. A two-electron reduced enzyme complex and an aldehyde product is formed by the reaction of the oxidized Cu^{2+} radical complex with the primary alcohol with constant k_{red}, during the first reductive semi-reaction (reaction 14). On the other hand, hydrogen peroxide is formed by the reaction of the reduced enzyme with O_2 with the k_{ox} constant, converting the active site of the metal-radical complex, during the second oxidative semi-reaction (reaction 15).

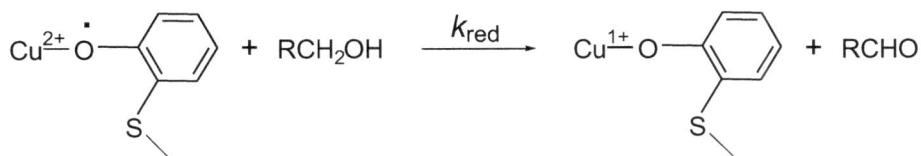

(14)

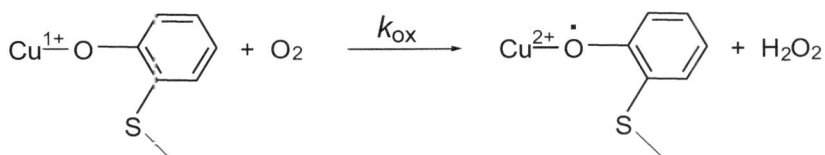

(15)

For the reaction of the enzyme with reduced substrates and with O_2 considering various substrates, the k_{red} and k_{ox} constants have been obtained. Additional evidence for the existence of a ping pong return mechanism is presented when the reoxidation rate is notoriously constant $k_{ox} = 0.98 - 1.02 \times 10^7$ $M^{-1}s^{-1}$, for substrates presenting a wide range of oxidation constants $k_{red} = 0.8 - 2.7 \times 10^4$ $M^{-1}s^{-1}$.

Relevant information on the catalytic constants of the reaction: K_m(galactose) = 175mM at saturating O_2 concentrations; $K_m(O_2) > 3$mM, is provided by kinetic experiments in the ground state. In the Michaelis complex due to the limited interactions between the alcohol substrate and the protein, the maximum K_m of the reduced substrate may represent an enzymatic strategy for a broad affinity substrate through binding affinity. The true binding affinity may be occluded by the linkage between the substrate and the energy of the active rearrangement site. As a product of higher participation in catalysis (reaction 16) a high affinity bond $K_D = k_2/k_1$, may be expressed by a large K_m value $K_m = (k_2 + k_3)/k_1$ [10].

(16)

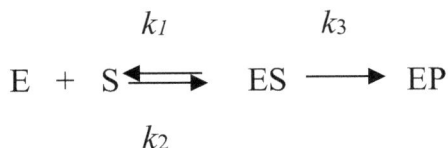

In reaction (16) k_1 is the entry constant, k_2 is the exit constant from binding to the substrate and k_3 means the concomitance of the catalytic reaction.

Under anaerobic conditions, a decrease in the intensity of the optical absorption of the enzyme in metal-radical complex reduction occurs when monitoring the reaction between the active enzyme and galactose. To ensure that the kinetics reflect substrate interactions and are not influenced by the environment, dioxygen can be excluded since the enzyme reacts rapidly with O_2. The reaction is biomolecular with a second order reaction constant of 1.58×10^4 $M^{-1}s^{-1}$, at low substrate concentrations, <10mM galactose. The reaction constant in the ground state reaction is K_m(galactose) = 175 mM, however, when the substrate concentration is raised, the constant exhibits saturation effects with K_m = 180 mM.

Based on the transformations that occur in the substrate and active site, substrate oxidation may be conceptually decomposed into discrete elements. Some of these transformations are: 1) the active site reduction of metal ions $Cu^{2+} \rightarrow Cu^{1+}$ -single electron transfer, SET; 2) the removal of the hydroxyl proton from the substrate in the conversion to the aldehyde product -proton transfer, HAT); 3) the removal of one of the substrate hydrogens Cα, one-atom abstraction; 4) the free radical reduction Tyr-Cys (Tyr-Cys$^{\cdot} \rightarrow$ Tyr-Cys) [10].

Acidification of the hydroxyl substrate in coordination with the Cu^{2+} ion in the active site facilitates proton transfer. A drastic effect on the pK of the hydroxyl has the coordination of a simple alcohol to a metal cation due to columbic and covalent effects.

By single electron transfer - SET in the inner sphere, the metal center is reduced. From the overlap between the valence orbitals of the coordinated substrate and the redox orbital in the Cu^{2+} ion depends the SET rate. A free radical alkoxyl is produced by the electronic oxidation of a coordinated alcohol.

For many substrates, this redox balance turns out to be unfavorable. The trend in two-electron oxidation potentials can parallel the trend in one-electron oxidation potentials. Thus, in the benzyl alcohol series, an initial SET step has a higher limiting rate for the 4-NO_2 derivative than for the 4-OCH_3 derivative. When an unfavorable initial SET is rate limiting in substrate oxidation, due to the decrease in C-H bonding in the alkoxy form, then a considerable effect on the isotope kinetics of the redox balance occurs.

Radicals with different structures are generated by both SET and HAT. An alkoxyl is the free radical formed by SET, which has an unpaired electron on the hydrogen atom (Scheme **6A**). While acetyl is the free radical formed by HAT which has an unpaired electron on the C_α of the substrate (Scheme **6B**). However, both processes can contribute to the determination of the catalysis constant.

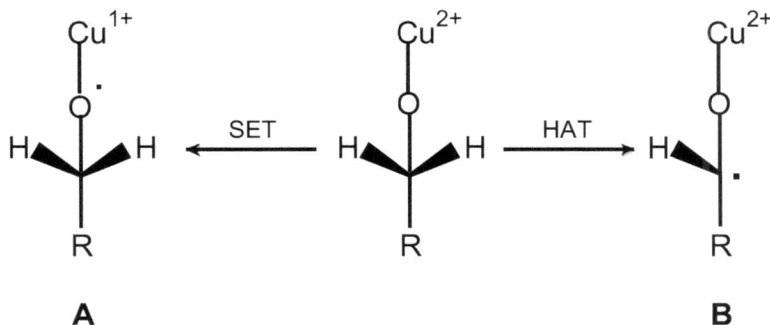

Scheme (6). Simple electron transfer (SET) and hydrogen atom transfer (HAT) mechanisms for the cuproxyl complex.

Considering the diffusion limit, the semi-reaction of oxygen reduction is very fast. *The enzyme may control the number of electrons derived from O_2 by preventing the escape of one-electron reduction products such as superoxide.* However, process inactivation results in the loss of superoxide at a rate of 2000-5000 returns.

There are two different forms of side coordination, (Fig. **12A**) and end coordination, (Fig. **12B**) by which the dioxygen is bound to the metal in the mononuclear copper complex. Strong covalent metal-ligand interactions characterize the lateral coordination of the *oxy* complex, such that the Cu^{3+}-peroxide bond contains a highly oxidized trivalent copper ion. The final bonding is favored by steric complexes, whereas the lateral complex bound to the oxygen requires two adjacent coordinated sites on the metal ion. The difference in the distribution of the valence electrons in these two classes of complexes presents different reactivity, with the lateral complex favoring the internal sphere reduction (center-metal) and the final complex favoring the external sphere reduction (center-oxygen).

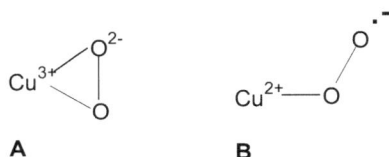

Fig. (12). Copper Mononuclear Coordination Complex. Explication in the text.

The crystallographically defined water molecule positions at the active site can provide an idea. A peroxide molecule also forms in other enzymes that react with oxygen, where the interaction patterns that stabilize the oxygen intermediate also stabilize a pair of water molecules in the absence of oxygen. The peroxide stabilization site is located between the two water molecules found in the active site of galactose oxidase and separated from each other by 3.2 Å [10].

Substrate oxidation reactions can be changed in the O_2 reduction (reaction 17). A Cu^{2+}-hydroperoxide adduct can be generated by atom transfer in the outer sphere (HAT) and reduction in the inner sphere (SET) to a dioxygen molecule. The return cycle could be completed by displacing the coordinated peroxide by protonation of the proximal coordinated oxygen atom.

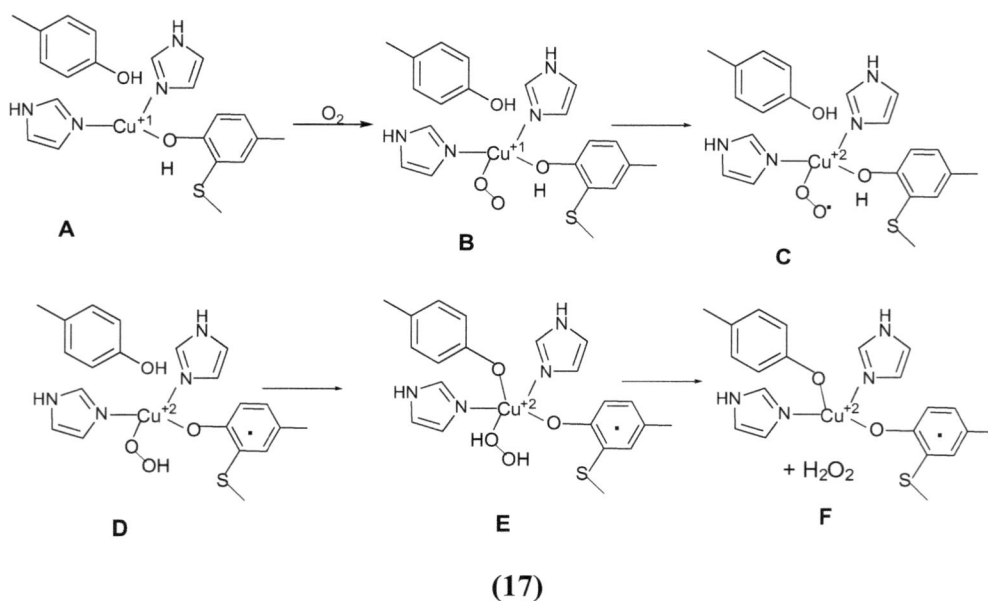

$$(17)$$

Self-processing reactions influence the oxidation state of the metal. The presence of a reduced metal ion is necessary for the reaction to proceed efficiently because the Cu^{1+} dependent rate $k_{xl} = 0.4$ s^{-1} is much higher than the Cu^{2+} dependent crosslinking rate $k_{xl} = 3.8 \times 10^{-5}$s^{-1}. As illustrated in reactions 18 and 19, based on the total electron count for the conversion of the Cu^{1+} or Cu^{2+} precursor complex into the active enzyme, the sensitivity of the crosslinking reaction to the oxidation state of the metal has been determined.

(18)

(19)

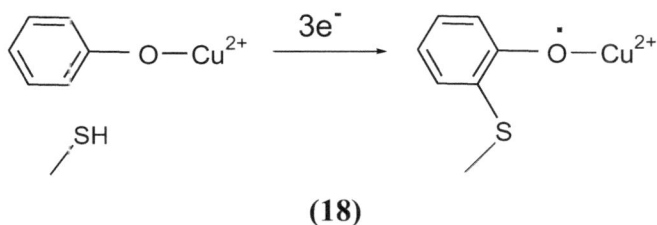

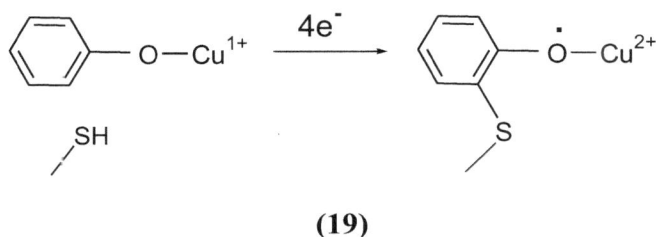

Tyr-Cys cofactor oxidation to the free radical and oxidative coupling of the tyrosine and cysteine side chains in a two-electron process are involved in the overall reaction. A reduction of the odd electrons of the dioxygen is necessary for a three-electron process without the presence of metal center oxidation. For dioxygen coupling it is more favorable that the redox center increases to a total four-electron reaction, as shown in reaction 20:

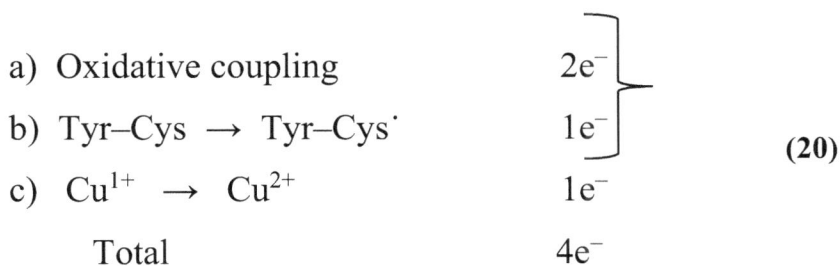

a) Oxidative coupling $2e^-$

b) Tyr–Cys $\rightarrow$ Tyr–Cys$^{\cdot}$ $1e^-$

c) Cu^{1+} $\rightarrow$ Cu^{2+} $1e^-$ **(20)**

        Total $4e^-$

Through Cu^{1+} reduction by a process adventitious from the sample the Cu^{2+} dependent process occurs which is consistent with the kinetic analysis of the cross-linking reaction. The biogenesis reaction involves free radical intermediates where the O_2 electron reduction is generated by oxygen reaction with the Cu^{1+} precursor. A mechanism consistent with experimental results has been proposed (reaction 21). The binding of Cu^{1+} to the preorganized site initiates the reaction (reactions 21 A and B). This complex reacts with dioxygen to generate an oxygenated species (Fig. **21 A** or **B**). The super*oxo* complex (reaction 21C) reacts with the thiol Cys228 in the outer sphere to form the thiyl free radical (reaction 21D). Addition of the thiyl radical breaks the aromatic ring conjugation (reaction 21E) in the Tyr272 ring system. This conjugation is restored on deprotonation and metal ion reduction (reaction 21F). A very fast reaction to form the oxidized metal-radical complex features the fully reduced active site (cross-linking of the complex) when reacted with a second O molecule [10].

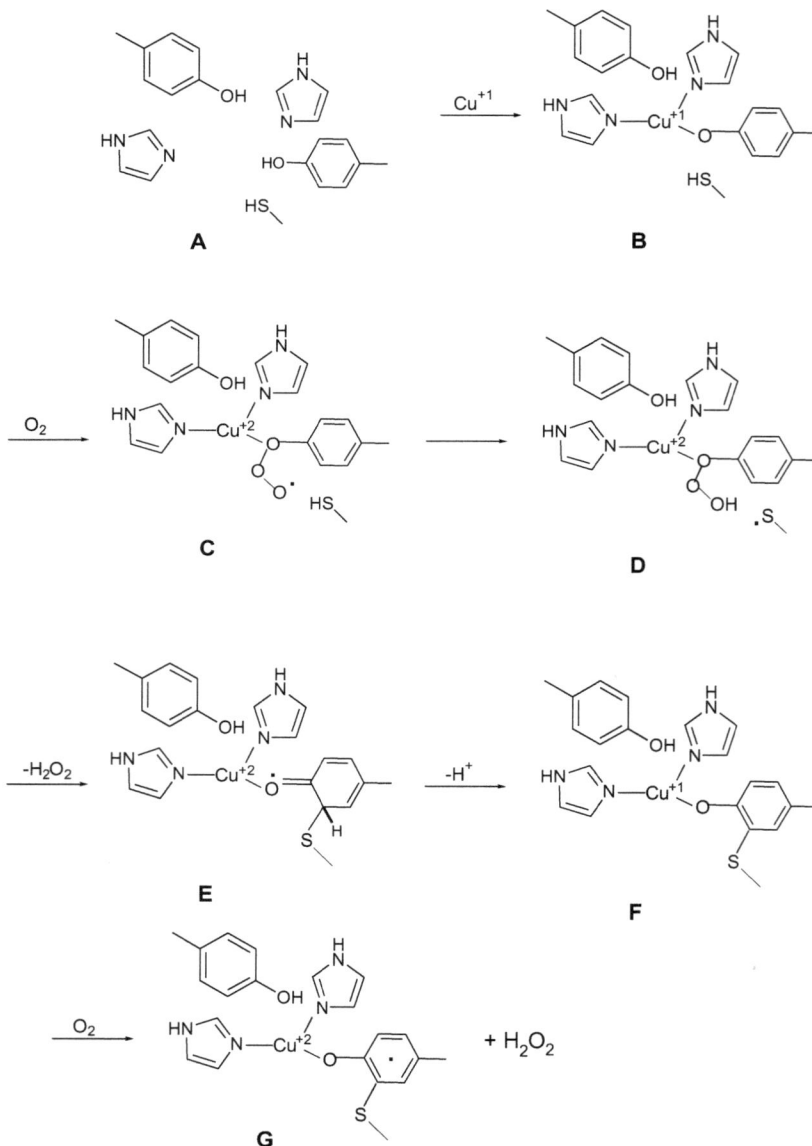

Reaction (**21, A, B, C, D, E, F, G**)

Furthermore, through the mechanism of the tyrosyl phenoxyl intermediate free radical (Fig. **13**) the lateral binding of the oxy species and the inner sphere reaction of the initial complex may be realized. The above mechanism also serves for the reaction dependent on the oxidation state of the metal, isotope sensitivity, O_2-stoichiometry, requirements for a metal-radical complex and pH.

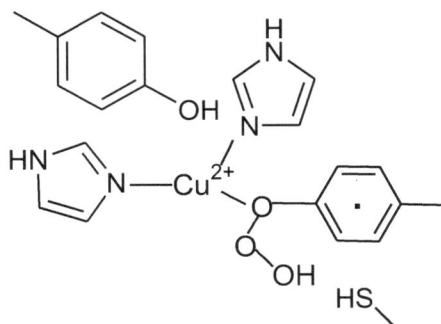

Fig. (13). Free radical tyrosyl phenoxyl intermediate.

1.2. CELLS AND ORGANELLES

1.2.1. Phagocytic Cells

The professional phagocytes essential for innate defense and necessary for rapid elimination of pathogenic microbes are neutrophils and monocytes. The NADH oxidase complex, which reduces molecular oxygen to microbicidal reactive oxygen species - microbicidal ROS, belongs to the NOX_2 antimicrobial system found in phagocytes. The NADH oxidase complex generates considerably higher levels of ROS than the other cellular oxidases, this complex is known as phagocytic oxidase.

In the cytoplasm, electrons are transferred from the active NADPH oxidase to molecular oxygen across the membrane. Then the superoxide anion O^-_2 formed by electron reduction on oxygen spontaneously dismutates to form hydrogen peroxide H_2O_2. More reactive radicals such as hypochlorous acid -HOCl or hydroxyl radical -·OH, are generated by the processing of the first ROS. An enzyme located in the azurophilic granules of neutrophils, the myeloperoxidase - MPO catalyzes HOCl formation, which is highly microbicidal.

This antioxidant is important in several pathologies because its excessive production causes damage to neighboring cells. Through the following reaction, superoxide stimulates the formation of radicals.

$$RNHCl + O^{\cdot-}_2 \rightarrow RNH^{\cdot} + Cl^- + O_2 \qquad (22)$$

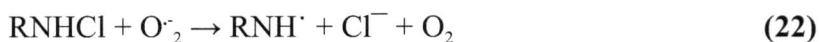

At sites of inflammation where HOCl is formed, the $O^{\cdot-}_2$ superoxide radical is generated by various cell types and pathways. Similarly, the superoxide radical $O^{\cdot-}_2$ is formed by the oxidative burst of phagocytes and is further formed by the following scavenging reactions: 1) of peroxyl radicals from aromatic side chains of proteins; 2) of α-hydroxyalkyl peroxyl radicals formed from sugars, Ser/Thr residues of proteins and polysaccharides, by one hydrogen atom abstraction; 3) from α-carbon peroxyl radicals from backbone proteins; 4) of α-amino-alkyl peroxyl radicals formed from Lys residues of proteins and amino sugars by one hydrogen atom abstraction; 5) from O_2 reaction with electrons from the electron transport chain and enzymes [11].

In neutrophils NADPH oxidase is assembled on membranes from the plasmalemma. Cytochrome b is 95% present in the membrane granules, so that NADPH oxidase is assembled and activated there. The remaining 5% of cytochrome b is found in the plasma membrane. The ROS produced by the oxidase assembled at the intracellular membrane are retained in a membrane-bound organelle -icROS. While the release of ROS to the extracellular medium - ecROS is produced by the oxidase assembled at the plasma membrane. The phagosome is the site of icROS formation; however, it has been observed to be formed by NADPH oxidase activation in cytochrome b-containing granules in the absence of the phagosome.

The phagocytosis process initiated by the internal folding of the plasma membrane giving rise to the membrane-enclosed phagosome constitutes the main defense of neutrophils against microorganisms. The mature phagolysosome formed by heterotypic fusion of granules also constitutes a form of phagosome processing. Thus, matrix components stored in the granules are carried to this phagolysosome. Cyt b is found in the phagolysosomal membrane only an instant after fusion of the phagosome with the Cyt b-rich granules, considering that the Cyt b located in the plasma membrane is now in the phagocytic vesicle and has been internalized with the prey.

In the phagolysosome, ROS generated by NADPH from azurophilic granules are processed by MPO. Also in the phagolysosome, target molecules are iodinated by the MPO-H_2O_2-halide system. However, it should be noted that ROS production is also generated in the compartment in the absence of the phagosome by stimuli that trigger NADPH assembly, although icROS formation is typical in the phagosome. ROS production in addition to the plasma membrane also occurs in organelles such as endosomes, nucleus, and endoplasmic reticulum, where NOX1-, 3-, 4- or 5-containing enzyme complexes mediate their activity [12].

The production of icROS also originates from receptor-ligand mechanisms activated by physiological and non-physiological stimuli. However, the entry of phagocytic particles into the phagosome is the main stimulus for icROS production. The phagocytic response in neutrophils is induced by complement receptors (CR1 and CR3), Fcγ receptors that recognize immunoglobulin opsonization and receptors that bind bacterial lectins, the glycoconjugate receptors. Thus, opzonized or lectin-binding particles are rapidly digested by the phagolysosome, leading to phagolysosome fusion and ROS production.

1.2.2. Peroxysomes

Oxidative stress and cancer have been demonstrated to be generated by peroxisome proliferators. Peroxisomes are organelles containing many oxidase enzymes that catalyze the reduction of divalent O_2 without the O^-_2 superoxide formation but have a high capacity to form H_2O_2 hydrogen peroxide. Some of these enzymes are aminoxidases, urate oxidase and glycolate oxidase. During β-oxidation, H_2O_2 is generated from fatty acid metabolism.

Although peroxisomes are generators of oxidative stress, they also contain an enzyme that reduces hydrogen peroxide to generate water and molecular oxygen, therefore the real contribution of these organelles to oxidative damage is unknown.

1.2.3. Mitochondria

Mitochondria are found in the cytoplasm of all eukaryotic cells. They are involved in some essential processes for cell function and survival such as: redox control, energy production, calcium homeostasis, certain biosynthetic and metabolic pathways. They contain two compartments bounded by the outer membrane and the inner membrane. For energy conservation and ATP synthesis, dependent on the high electrochemical gradient originating from the electron transport chain, the mitochondrial inner membrane controls its permeability. Whereas the mitochondrial outer membrane is permeable to small metabolites.

Oxidative phosphorylation for cellular energy generation is the main function of mitochondria. The coupling of electron transport by the electron transport chain - ETC, across the mitochondrial inner membrane and ATP formation by F_1F_0-ATP synthase, to the proton pump is performed by mitochondria through oxidative phosphorylation. Structurally the ETC is constituted by five complexes designed for energy production, which involve approx. 80 protein components. These complexes are in order: Complex I -NADH dehydrogenase, complex II -succinate dehydrogenase, complex III -ubiquinone cytochrome c oxidoreductase, complex IV -cytochrome oxidase and complex V -F_1F_0-ATP synthase.

In complex I or complex II, the electron pairs enter and in complex IV they leave. Through ubiquinones or cytochrome c the electrons are transmitted between the complexes. One pair of electrons is given up by NADH to FMN in complex I. At the Fe/S_{II} center, the electrons leave complex I and pass to the coenzyme Q_{10} ubiquinone. In Fig. (**14**), the reduction potentials of the different resting states with their metabolic pathway are presented. Between the $(FeS)_{III}$ and $(FeS)_{II}$ centers there is a decrease in the potential of 225 mV that is coupled to the phosphorylation reaction. However, the reduction potential is 65 mV for each of the other resting centers of complex I [8].

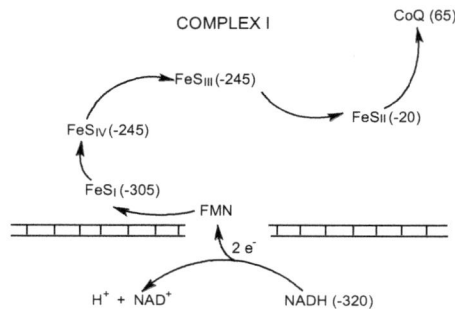

Fig. (14). Transport Chain Reactions in the Mitochondrial Complex I. Numbers in parentheses are the reduction potentials.

Cofactors such as NADH and $FADH_2$ which are reduced hydrogen transporters are produced by the oxidation of molecules such as carbohydrates, proteins, and lipids through cellular metabolism. To the electron transport chain -ETC, these cofactors donate their electrons. A redox potential that is present across the chain directs the movement of electrons between the components of the ETC. As electrons flow through the respiratory chain complexes I, III and IV pump protons across the inner membrane, this produces the proton driving force Δp, which is an electrochemical gradient (membrane potential) and a small chemical gradient (pH difference), caused by differences in the electrochemical potential across the mitochondrial inner membrane. As protons are transported back from the intermembrane space to the mitochondrial matrix, the energy conserved in the proton gradient across the mitochondrial inner membrane is utilized by complex V to generate ATP from ADP and Pi. In the last step of ETC, molecular oxygen is the final acceptor of electrons that are reduced to water by complex IV. Thus, in respiration, the process of substrate oxidation and oxygen reduction is coupled to ATP formation [13].

Through the inner membrane proton conductance pathway that avoids ATPasa, protons pumped out of the matrix can be returned to the mitochondria, utilizing some of the energy available in the electrochemical gradient, since not all the energy is coupled to ATP synthesis. The energy that is dissipated and released as heat comes from the oxidation of metabolic fuel (Fig. **15**). Through transporter proteins, substrates enter from the cytosol into the mitochondrial matrix. In the matrix, the free energy released by the oxidation of these substrates is conserved in the reduced NADH and FADH$_2$ molecules. Through the ETC electrons are transferred from NADH and FADH$_2$ to oxygen. A negative proton force -Δp inside the matrix is caused by proton pump conduction across the mitochondrial inner membrane. As protons are transported back from the intermembrane space to the mitochondrial matrix *via* F$_1$F$_0$-ATP, the conserved energy in the proton gradient leads to the synthesis of ATP from ADP and Pi.

20-25% of the basal metabolic rate corresponds to non-productive proton release, which is physiologically important. The functions of proton release are attenuation of ROS production, regulation of energy metabolism or carbon fluxes, thermogenesis, and control of body mass. In endothermic and exothermic animals, basal proton conductance occurs in isolated mitochondria, while the sterile proton cycle is thermogenic.

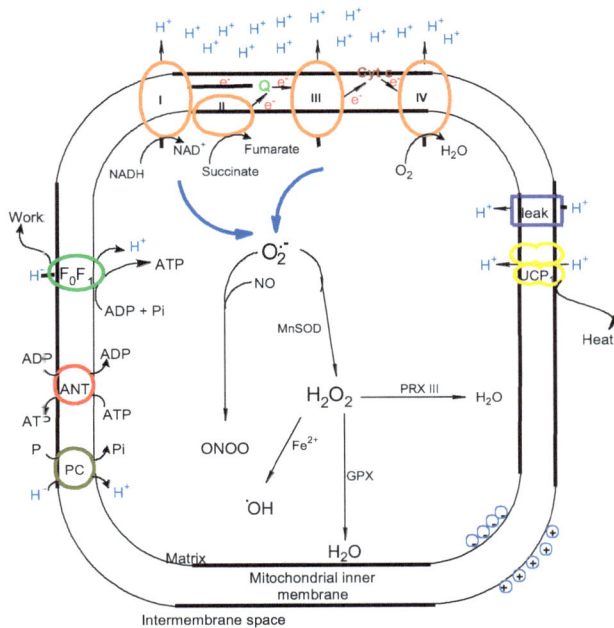

Fig. (15). Reactive oxygen species -ROS production in the mitochondrial electron transport chain -ETC.

The mitochondrial respiratory chain -ETC is the main site of non-enzymatic ROS production in cells. As mentioned in previous paragraphs, the ETC contains several redox centers capable of transferring an electron to oxygen to form $O_2^{\cdot-}$. Complexes I on the matrix side of the mitochondrial membrane and III in the matrix and intermembrane space are the main sites to produce $O_2^{\cdot-}$ superoxide. In complex I the oxygen reductant for superoxide production is unknown. However, electron transporter sites such as flavin, semiquinones and Fe/S center may generate $O_2^{\cdot-}$ superoxide. For its part, the unstable molecule ubisemiquinone located in complex III is the main oxygen reductant to produce superoxide, according to the Q-cycle mechanism of the complex. Tissue type, experimental conditions and species affect the contributions of complexes I and III to reactive oxygen species-ROS production. Complex IV produces little or no superoxide.

The metabolic states of mitochondria considerably affect the rate of ROS production, in addition to the effect exerted by inhibitors. Superoxide production is highest in state 4 metabolic conditions when oxygen consumption is low and proton motive force is high, therefore, the electron transport chain complexes - ETC are in a reduced state. This occurs when ATP demand is low and proton motive force is high. In addition, superoxide production decreases significantly at ADP addition and phase 3 transition of respiration. The highest mitochondrial ROS production occurs in state 4 conditions because in non-phosphorylated conditions the proportion of consumed oxygen resulting as ROS is higher than in phosphorylated conditions. In addition to the above, in intact cells, mitochondria normally breathe between states 3 and 4.

In the mitochondrial inner membrane, cytochrome *c* -Cyt *c* (section 1.1.1. chapter 3) is found in high concentrations of 0.5-5mM. Cyt *c* represents the coming and going of cytochrome *c* reductase complex III and cytochrome *c* oxidase complex IV during mitochondrial respiration. The associated Cyt *c* prevents oxidative damage by participating in electron transport and superoxide scavenging. While in peroxidase activity, closely related Cyt *c* participates [13]. Complex IV is protected by oxidized cytochrome *c*, the Fe^{3+}-Cyt *c* from oxidative damage caused by H_2O_2. The Cyt *c* interacts with cardiolipin, altering its tertiary structure and increasing its peroxidase activity.

Velayutham *et al.*, 2011 [14], employing UV-visible absorption spectroscopy technique and electron paramagnetic resonance (EPR), investigated the role of Fe^{3+}- Cyt *c* in both superoxide radical formation from NADH oxidation and H_2O_2 scavenging. They found that Fe^{3+}- Cyt *c* in H_2O_2 detoxification acts as a peroxidase forming a type I intermediate compound. Therefore, Fe^{3+}- Cyt *c* oxidizes endogenous antioxidants such as GSH, ascorbate and NADH in the presence of H_2O_2 (Scheme 7) .

$$Fe^{3+} \text{ Cyt } c + H_2O_2 \rightarrow \underset{\text{Compound I}}{O=Fe^{4+} \text{ Cyt } {}^{\cdot+}c} + H_2O$$

$$\downarrow \text{ DMPO, GSH, AscH}^-, \text{NADH}$$

$$\text{DMPOX, GS}^{\cdot}, \text{Asc}^{\cdot-}, \text{NAD}^{\cdot}$$

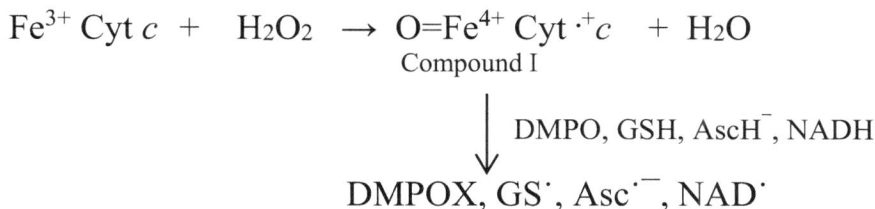

Scheme (7). Activation mechanism of Fe^{3+}- Cyt c to a type I intermediate compound and oxidation of its substrates.

Through the peroxidase activity of Fe^{3+}- Cyt c, oxidation of NADH occurs generating $NAD^{\cdot}$, which in turn reacts with molecular oxygen to form NAD^+ and superoxide radical. A second order reaction constant of 1.9×10^9 $M^{-1}s^{-1}$ exists between $NAD^{\cdot}$ and O_2 (Scheme 8).

$$Fe^{3+} \text{ Cyt } c + H_2O_2 \rightarrow \underset{\text{Compound I}}{O=Fe^{4+} \text{ Cyt}^{\cdot+}c} + H_2O$$

$$\downarrow \text{ NADH}$$

$$NAD^+ + O^{\cdot}_2 \overset{O_2}{\longleftarrow} NAD^{\cdot}$$

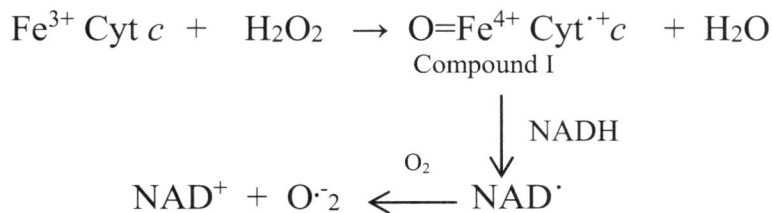

Scheme (8). Oxidation mechanisms of NADH by intermediate compounds type I with the generation of the superoxide radical.

1.2.4. Microsomes

Microsomal electronic transport systems also produce $O^{\cdot}_2$ and H_2O_2. When microsomal fractions are incubated with NADPH, oxygen-free radicals are produced at a faster velocity as the oxygen concentration increases.

The cytochrome P-450 is a group of *hemic* proteins located in the endoplasmic reticulum and involved in xenobiotic metabolism. Their function is to oxidize substrates at the expense of O_2; an oxygen atom binds to the substrate and other forms water. NADPH donates the electrons required by cytochrome P-450 through a flavoprotein called NADPH-cytochrome P-450 reductase. Oxygen-free radicals in this system are produced in two ways: by autooxidation of the NADPH-cytochrome P-450 reductase or by decoupling of the catalytic cycle from the P-450. The decoupling of the normal redox cycle of these cytochromes is induced by various compounds and causes an electron flow deviation to O_2 to produce $O^{\cdot}_2$, rather than reducing the substrate [1].

Cyt P-450-dependent monooxygenases catalyze the reductive cleavage of molecular oxygen O_2. One of the two oxygen atoms is released as water, while the other is transferred to the substrate. A FAD-containing auxiliary enzyme of the connexin NADPH + H^+ transfers the reducing equivalents to the monooxygenase (Fig. **16**).

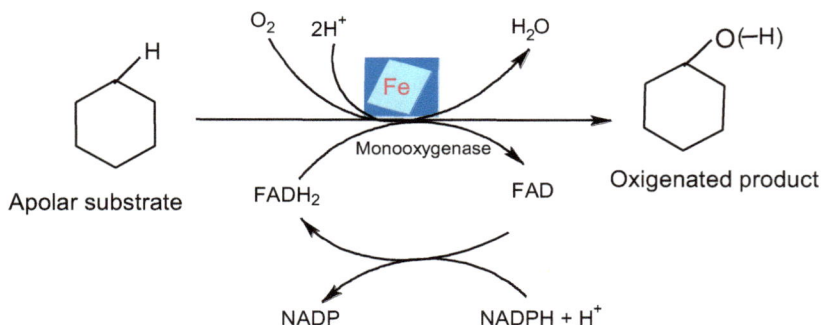

Fig. (16). Monoxygenase activity dependent on Cyt P-450.

In the reaction mechanism of the P-450 catalysis, in the steady-state *heme* iron is trivalent. The substrate initially binds near the *heme* group (reaction 23) where one-electron transfer from $FADH_2$ reduces the iron into the divalent form to which an O_2 molecule can be bound (reaction 24). O_2 bond reduction to peroxide occurs by the transfer of a second electron and a change in the valence of iron (reaction 25). The hydroxyl ion of this intermediate is fragmented, and proton conductance generates H_2O and activated oxygen (reaction 26). In this ferric radical, iron is tetravalent. An OH group is formed because the activated oxygen atom is attached to the C-H bond of the substrate (reaction 27). The product dissociation returns the enzyme to its initial state (reaction 28) (Fig. **17**) [15].

The microsomal desaturant system responsible for the introduction of double bonds in fatty acids contains a flavoprotein which is a reductase, a desaturase, and a cytochrome b_5. This system also needs NADH or NADPH and O_2 to oxidize its substrates. The electrons are transferred by cytochrome b_5 reductase from NADH or NADPH to cytochrome b_5 and from there to desaturase, which oxidizes fatty acid with O_2 and forms water. The cytochrome b_5 and flavoprotein can give electrons to molecular oxygen and form O^-_2 [1].

The flavin cytochrome b_5 reductase enzymes that have FAD as a prosthetic group and lack heavy metals are NADPH cytochrome *c* oxidoreductase and NADH cytochrome b_5 oxidoreductase. The reduction of dichlorophenolindophenol and ferricyanidin is catalyzed by cytochrome b_5 reductase and cytochrome b_5. In turn,

NADH-cytochrome c reductase also catalyzes the reduction of neotetrazolium, ferricyanidin, various quinones and dichlorophenolindophenol. Therefore, these enzymes catalyze one-electron acceptor reduction as well as two-electron acceptor reduction [16].

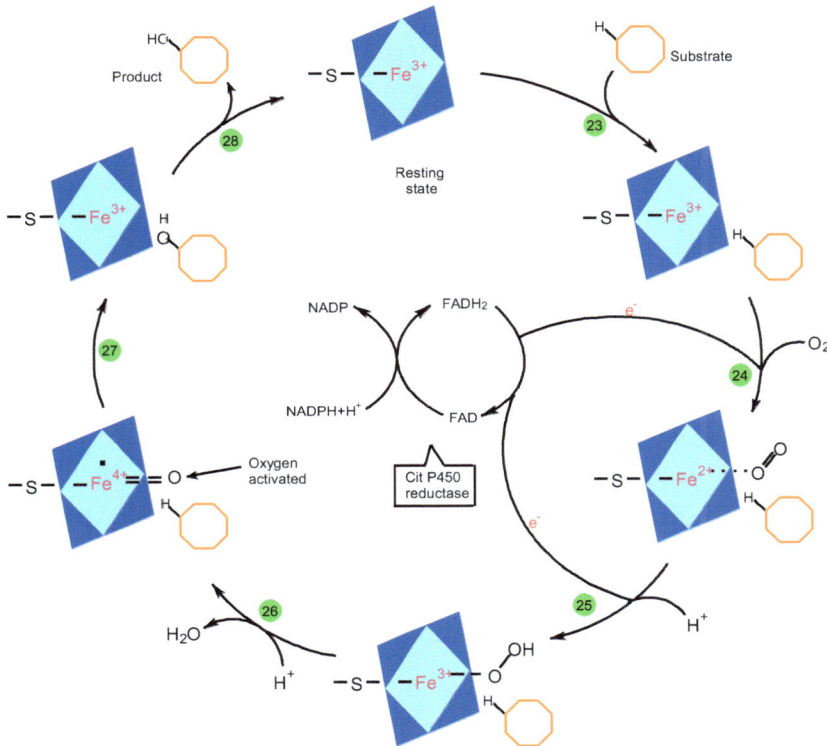

Fig. (17). Catalytic cycle of the P-450. Explanation of the reactions in the text.

Since ferricyanidin and cytochromes are electron acceptors for enzymes, in microsomes the flavin-reduced enzymes are oxidized in two-step reactions. This mechanism involves during the reaction an oxidized intermediate in the flavin enzymes. For ferricyanidin transport, the general reaction involves the cytochrome c reduction and the processing of 2 FADH $\leftrightarrow$ 2 FADH$_2$, considering that NADP-cytochrome c reductase contains per mole of enzyme, 2 moles of FAD.

When the reduced flavin enzyme H$_2$F reacts with electron acceptor A, the following reaction is generated:

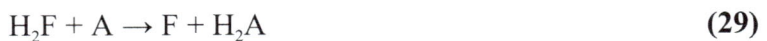

$$H_2F + A \rightarrow F + H_2A \tag{29}$$

The quinone molecule receives an electron from the flavin enzymes and produces a semiquinone which is free of enzymes, is explained by the following mechanism (Scheme **9**).

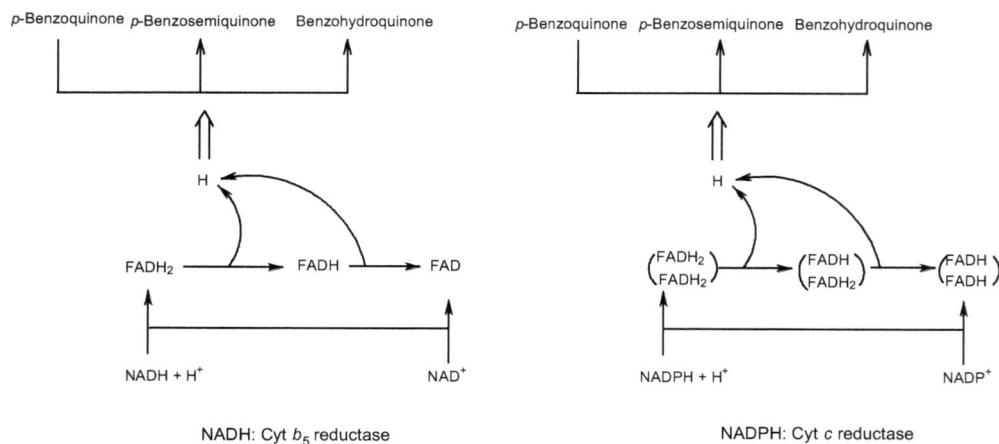

Scheme 9. Quinone reduction mechanism by Microsomal Flavin Enzyme.

The reaction (29) is replaced by the following reaction, when in the reaction between a flavin enzyme and two acceptors an electron transfer mechanism is established.

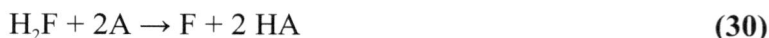

$$H_2F + 2A \rightarrow F + 2\,HA \tag{30}$$

The semiquinone in the electron transport system acts as an electron donor and is much more reactive than the corresponding quinone in its reduced form. The electron acceptors for the semiquinone are molecular oxygen and cytochromes, so that reaction with these electron acceptors and dismutation are the main pathways of semiquinone disintegration.

Monodehydroascorbate-stimulated NADH oxidation in microsomes may be explained by electron flow from cytochrome b_5 reductase to monodehydroascorbate. The above occurs in the electron transfer pathways of microsomal flavin enzymes. However, there is also electron flow from cytochrome b_5 to monodehydroascorbate.

The reduction of two-electron acceptors, including molecular oxygen can be catalyzed by many flavin enzymes (Scheme **10**) [16].

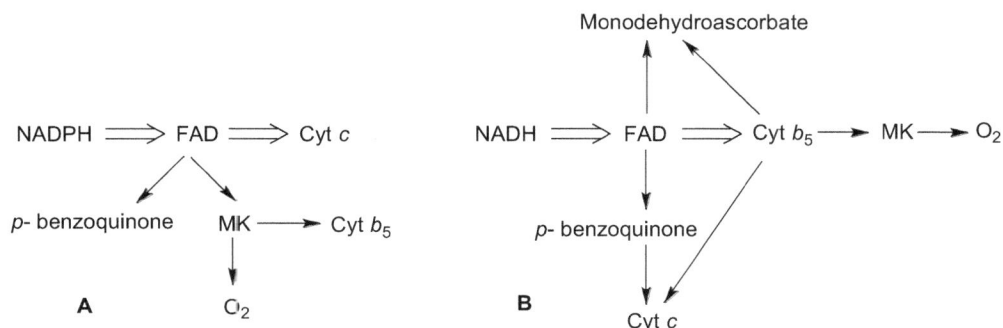

Scheme (10). One-electron transfer mediated by the two-electron system in flavin microsomal enzymes. **(A)** Cytochrome b_5 reductase. **(B)** NADPH-cytochrome c reductase.

The one-electron reduction of quinones is catalyzed by microsomal flavoproteins such as NADPH-cytochrome c reductase and cytochrome b_5 reductase. The one and two electron reduction of p-benzoquinone or oxygen is catalyzed by xanthine oxidase. The above mechanisms depend on the acceptor concentration.

For NADH, cytochrome c is a very slow electron acceptor. The addition of p-benzoquinone stimulates the reduction of cytochrome c, which in turn cannot be reduced by NAD(P)H dehydrogenase.

This enzyme in turn catalyzes quinone and pigment reduction. It may also catalyze the cytochrome c reduction but in the presence of naphthoquinone derivatives, not in the presence of benzoquinone derivatives. Mammalian quinone reductase, on the other hand, reduces MK. Under aerobic conditions, the MKH_2 autooxidation limits the rate of reaction, such that NADPH is equivalent to the amount of MK added. The above results suggest that in the presence of NAD(P)H dehydrogenase, quinone-mediated electron transfer to cytochromes and oxygen is slow [17].

Soluble NADH dehydrogenase catalyzes quinone reduction such as MK and ubiquinones because of its diaphorase activity. Therefore, most flavoproteins catalyze the transfer of electrons in molecular oxygen, quinones and pigments.

If the enzyme catalyzes the transfer of one or two electrons to these acceptors, it is important to know. To distinguish the two mechanisms, the parameter κ has been introduced:

$$K = \frac{2k_d(\text{semiquinone})_s^2}{v} \qquad \textit{(Equation 7)}$$

where k_d is semiquinone dismutation constant and v is the quinone reduction rate in the ground state.

Measuring the semiquinone concentration in its ground state, the value of K may be estimated by the ESR technique. This value of κ is between 2 and 0. For the two-electron transfer mechanism the value of κ is 0, and for a typical electron transfer mechanism the value of κ is 2. For example, in the reaction between quinone and quinol, to avoid the non-enzymatic quinone formation, the experimental conditions should be selected by reducing the pH of the solution and acceptor concentration, especially when it is desired to measure κ according to equation 7. In this equation, only when k_d is known may the value of κ be estimated and during the reaction failures in the ESR signals do not mean that κ = 0. Therefore, when k is notoriously large or the reaction rate v is low, the ESR method is not applicable.

When benzoquinone is an acceptor, the ESR method is very useful for quantitative analysis of the mechanisms, however, this method is not useful when MK or molecular oxygen are acceptors, because the dismutation constants for MKH and O_2^- are high. For the analysis of the electron transfer mechanism, the three reaction systems presented in the following figure may be utilized. Experimental conditions may be established for electron flow from a flavoprotein to a final acceptor only when this flavoprotein catalyzes the electron reduction of the transporter molecule. Thus, κ may be measured as the ratio between the rate of cytochrome reduction and the rate of quinone or O_2 disappearance in the absence of cytochromes. The rate of reduction of cytochromes under optimal conditions [17].

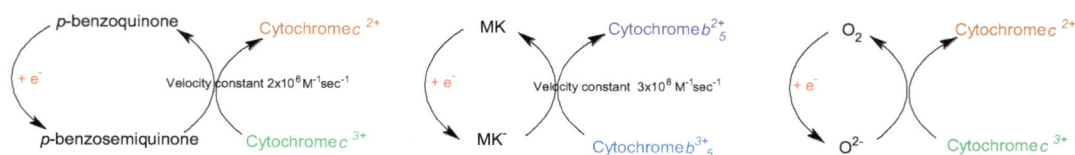

Fig. (18). Useful reaction systems for electron transfer mechanisms.

The estimation of κ, when cytochromes are utilized as final acceptors, could imply that: 1) the reduction of cytochromes by benzohydroquinone or MKH_2, could be avoided with a low pH in the reaction solution, 2) the disappearance of semiquinones or O_2^- by dismutation may be scarce when the formation of semiquinones or O_2^- is slow and there is a high cytochrome concentration.

According to the way of transferring electrons to the acceptors, flavoproteins are divided into three groups: 1) The one-electron reduction of the acceptors forming free radicals of two-electron acceptors $\kappa = 2$, is catalyzed by flavoproteins-1. 2) The reduction by two electrons of quinones or oxygen, the main products being in the form of two reduced equivalents of the acceptor molecule $K = 0$, is catalyzed by the flavoproteins-2 and 3) When K is between 0 and 2 a mixed mechanism and is catalyzed by another group of flavoproteins, the flavoproteins-1,2.

Microsomal NADPH cytochrome c reductase catalyzes oxygen-dependent NADPH oxidation in the presence of MK, but not in the presence of MKH_2. Xanthine oxidase catalyzes the cytochrome c reduction mainly in the presence of oxygen. Aldehyde oxidase has properties like those of xanthine oxidase. In the reaction of dihydroorotate dehydrogenase, the relationship between the reduction of cytochrome c and the hydrogen peroxide formation, in a mechanism involving an oxygen free radical, has been quantitatively investigated.

Succinic dehydrogenase catalyzes the reduction of phenazinametosulfate by a mechanism involving an electron transfer mechanism. Thus, in the presence of phenazinametosulfate as a transporter, the enzyme also catalyzes the cytochrome c reduction. To the flavoproteins-2, belong a group of flavoproteins called oxidases: D-amino acid oxidase, L-amino acid oxidase, lactate oxidase, glucose oxidase, and glycolate oxidase.

Many flavoprotein dehydrogenases, such as NADPH-cytochrome c reductase, NADPH-cytochrome c_2 reductase, NADH-cytochrome b_5 reductase, NADPH-cytochrome f reductase and succinic dehydrogenase, react with one-electron acceptors more rapidly than with two-electron acceptors. However, they have diaphorase activity. Regarding some properties of flavoproteins, in the case of dehydrogenases the blue or neutral form is the stable form produced in a partial reaction, while with few exceptions, the semiquinoid form of oxidases is the red or anionic form.

When quinones are in an electron transport system, they act as a one-electron transporter and not as a two-electron transporter. Considering the above, the following reactions may occur when the two-electron acceptor A reacts with the reduced flavoprotein H_2FP.

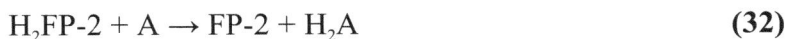

$$H_2FP\text{-}1 + 2A \rightarrow FP\text{-}1 + 2\,HA \qquad\qquad (31)$$

$$H_2FP\text{-}2 + A \rightarrow FP\text{-}2 + H_2A \qquad\qquad (32)$$

In the xanthine oxidase reaction from the reduced form of flavin to oxygen, a direct electron transfer is required for the formation of O_2^-. In the diaphorase activities of soluble NADH dehydrogenase, the sulfur iron system is not involved. Therefore, it is important to emphasize that microsomal flavoproteins may catalyze the typical electron transfer without the presence of metal ions. Thus, for the flavoprotein catalyzes electron transfer to acceptors, metal ions are not indispensable [17].

1.3. Cytosolic Molecules

The autooxidation of small cytosolic molecules can produce oxygen free radicals. Some examples of these molecules are catecholamines, flavins, tetrahydropterins and quinones or thiols and diphenols. As a univalent oxygen reduction, O_2^{-} is produced in all cases. Furthermore, if the original molecule is regenerated by reducing agents, a non-enzymatic redox cycle is established. This auto-oxidation process begins or accelerates in the presence of transition metals.

2. EXOGENOUS SOURCES

The frequent exposure to different forms of metal contamination from environmental composition and/or occupational activities generates alterations in the normal functioning of living cells. Thus, metal working, mining, hydrocarbon exploitation, leather tanning and groundwater contamination, among others, are some of the main sources of exposure to metal contamination. These metals include chromium (Cr), cadmium (Cd), vanadium (Va), mercury (Hg), nickel (Ni), zinc (Zn), lead (Pb), cobalt (Co), arsenic (As) and beryllium (Be), among others. Intense exposure to certain metals causes damage to cells and tissues, although small traces of certain metals such as iron are necessary for the normal functioning of living organisms.

Transition metal toxicity varies according to the chemical characteristics of each metal. Thus, the specific differences of each transition metal are related to solubility, absorption, transport, chemical reactivity, and complexes formed in the organism being the basis for the difference in toxicity that these metals present. ROS production by metals seems to have a common mechanism, however, each metal has its own action mechanism.

The *Fenton*-type reaction is one of the main mechanisms for the metal-mediated generation of oxygen free radicals. In this reaction ·OH radicals and an oxidized metal ion are formed from the reaction of H_2O_2 hydrogen peroxide with a transition metal:

$$\text{metal}^{n+} + H_2O_2 \rightarrow \text{metal}^{n+1} + \text{·OH} + \text{OH}^{-} \tag{33}$$

Fenton type reagents include metals such as iron, chromium (III), (IV) and (V), copper, nickel (II), cobalt (II) and vanadium (V). These ions produce $\cdot$OH, however, the production efficiency varies considerably. For example, due to the high potential that can be modified by their own chelation, the efficiencies for cobalt (II) and nickel (II) are very low. These metal ions react with H_2O_2 and lipid peroxides to produce hydroxyl $\cdot$OH radicals and lipid radicals in the presence of chelating agents such as Gly-Gly-His or thiol-containing agents. Also $\cdot$OH radicals are readily produced by the reaction of chromium(VI) metal. However, glutathione and sulfhydryl-linked protein diminution occurs in all toxic forms in the reaction of metal ions such as nickel, lead, mercury, and cadmium.

$$M(X) + O^{\cdot}_2 \rightarrow M(X-1) + O_2 \tag{34}$$

$$2O^{\cdot}_2 + 2H^+ \rightarrow H_2O_2 + O_2 \tag{35}$$

$$M(X-1) + H_2O_2 \rightarrow M(X) + \cdot OH + OH^- \tag{36}$$

In membranous organelles such as mitochondria, peroxisomes and microsomes, Fenton-type reactions are commonly performed. Thus, the production of reactive oxygen species-ROS can be initiated by some transition metal by more than one mechanism, involving more than one type of compartment or cell.

2.1. *Haber – Weiss* **Reactions**

Metal-mediated free radicals are also produced through *Haber-Weiss* reactions. In these reactions $\cdot$OH radicals are generated by the reaction of H_2O_2 with an oxidized metal ion that has been reduced by $O^{\cdot}_2$:

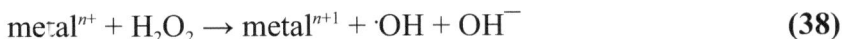

$$metal^{n+1} + O^{\cdot}_2 \rightarrow metal^{n+} + O_2 \tag{37}$$

$$metal^{n+} + H_2O_2 \rightarrow metal^{n+1} + \cdot OH + OH^- \tag{38}$$

(overall)

$$O^{\cdot}_2 + H_2O_2 \xrightarrow{\quad metal^{n+1}/metal^{n+} \quad} \cdot OH + O_2 + OH^- \tag{39}$$

When macrophages and cellular components during the respiratory burst generate large amounts of $O^{\cdot}_2$ the mechanisms of $\cdot$OH radical generation of the Haber-Weiss type become very important especially during phagocytosis. The respiratory burst is triggered by phagocytes when they trap metal-containing particles. For the Haber-Weiss reaction to occur, the presence of a metal ion is required, as the conversion of $O^{\cdot-}_2$ to H_2O_2 is too slow to be physiologically important. Thus, in a limited number of available metals a large amount of OH radicals are generated [18].

The Haber-Weiss type reaction can involve metals such as chromium (III), (IV), (V) and (VI), vanadium (V) and cobalt (I). The chromium (VI) mediated $^{\cdot}$OH generation reaction is represented in the following reaction scheme:

$$Cr(VI) + O^{\cdot-}_2 \rightarrow Cr(V) + O_2 \qquad\qquad (40)$$

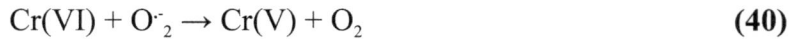

$$Cr(V) + H_2O_2 \rightarrow Cr(VI) + {^{\cdot}OH} + OH^- \qquad\qquad (41)$$

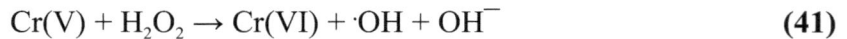

(overall)

$$O^{\cdot-}_2 + H_2O_2 \xrightarrow{\quad Cr(VI)/Cr(V)\quad} {^{\cdot}OH} + O_2 + OH^- \qquad (42)$$

2.2. Iron

Iron is essential for almost all living organisms; it is the catalytic center of iron-dependent enzymes and integrates the *heme* fraction. More than 1000 iron-dependent proteins have been identified in eukaryotes. Iron-sulfur groups, 40% of non-heme proteins, and heme cofactors are some of the sites where iron as a metal ion is found in proteins. The *heme* group is found in cytochromes, myoglobin, and hemoglobin -Hb.

Human adults have approximately 3 to 5 g of iron in the body, absorb about 1 to 2 mg of iron per day; thus, approximately 36 mg is the daily replenishment of iron in the plasma. In the Hb present in erythroid cells, 65-75% of this iron is found. Mainly in the liver in the form of ferritin, 10-20% is stored, and in the *heme* bond of striated muscle myoglobin, 3-4% is found. For *heme* biosynthesis alone, immature red blood cells consume approx. 25 mg of iron per day.

The ferrous ion Fe^{2+} is spontaneously oxidized at physiological pH to produce reactive oxygen species -ROS, such as $O^{\cdot-}_2$ superoxide, $^{\cdot}$OH hydroxyl radicals and H_2O_2 hydrogen peroxide. The Fenton reaction is the typical example of the prooxidant activity of iron, in which hydrogen peroxide reacts with iron to produce hydroxyl radicals according to the following reaction:

$$Fe^{2+} + H_2O_2 \rightarrow Fe^{3+} + OH^- + {^{\cdot}OH} \qquad\qquad (43)$$

The well-known intracellular redox iron store - LIP is the source of iron for the Fenton reaction. However, LIP is small and is not a major source of ROS in many organisms. In LIP iron is in the ferric form, complexed with citrate or ADP consisting of cytotoxic ferrous iron. LIP is retained only as a small fraction 3-5% of the total cellular iron, due to its toxicity. Therefore, that LIP levels demonstrate the general state of iron in the cell. Thus, in the plasma of some pathologies catalytic iron for Fenton reactions is found in the micromolar range but is not found in healthy people. Excessive production of hydroxyl radicals and lipids occurs as a direct consequence of excess Fe^{2+}, when the iron concentration exceeds the iron detoxification capacity, *i.e.* trapping, transport and oxidation of Fe^{2+} iron to less reactive Fe^{3+}. Similarly, oxidative damage is caused by iron deficiency due to increased electron release from iron-dependent mitochondrial ETC components [19].

Iron transfers an electron to dioxygen producing superoxide O^{-}_2 radicals, thus catalyzing the formation of oxyradicals since O^{-}_2 is the precursor of H_2O_2. This in turn reacts with iron (II) to generate the highly reactive ·OH hydroxyl radical through the Fenton reaction. The O^{-}_2 superoxide reacts rapidly with 4Fe-4--containing proteins, whereas the ·OH hydroxyl radical reacts with lipids and proteins. To react with proteins containing 4Fe-4S groups, the O^{-}_2 must diffuse over long distances, which results in increased levels of H_2O_2 and free iron, which are components of the Fenton reaction. As a result of this specific reaction, hydroxyl ·OH radicals are generated, which rapidly attack neighboring molecules [20].

The catalytic activity of iron or its complexes involves redox reactions between oxygen and biological molecules that generate reactive oxygen species - ROS (section 1.2.3. Chapter 1). ROS are also formed by the reaction of iron with histidine, citrate, 5'-diphosphate ADP and ethylenediaminetetraacetic acid-EDTA and other chelating agents.

Through the Fenton reaction the chelated iron acts as a catalyst, facilitating the formation of ·OH hydroxyl radicals by the reaction of superoxide with hydrogen peroxide. The ·OH hydroxyl initiates lipid peroxidation (reactions 43 and 44).

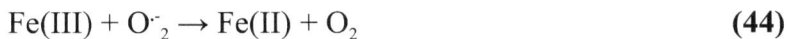

$$Fe(III) + O^{-}_2 \rightarrow Fe(II) + O_2 \qquad \textbf{(44)}$$

Hydrogen peroxide and superoxide anion, which are formed by xanthine oxidase, participate in the iron redox reactions that facilitate the oxidation of Fe(II) or the reduction of Fe(III), thus collaborating in iron-catalyzed lipid peroxidation.

2.2.1. Iron Storage Proteins

2.2.1.1. Transferrin

In vertebrate serum, transferrin-Tf is the major iron-binding protein. It is a glycoprotein of 698 amino acids, produced from liver cells into plasma. In the presence of bicarbonate anion, it binds two Fe^{3+} iron atoms. Apotransferrin is the iron-free Tf. In Tf the iron binding sites are in the highly conserved N-terminal and C-terminal sites. Iron binds to two tyrosine residues, an aspartic acid, and a histidine.

2.2.1.2. Ferritin

In prokaryotes and eukaryotes Tf is the main intracellular iron storage protein. Iron storage involves the following processes: conduction of Fe^{2+} to the envelope protein, in the center of the dinuclear ferroxidase the oxidation of Fe^{3+} by molecular oxygen, the deposition of iron as ferrihydrite in the cavity, followed by its mobilization as Fe^{3+} into this cavity.

Ferritin catalyzes iron oxidation in the next reaction:

$$4Fe^{2+} + 4H^+ + O_2 \leftrightarrow 4Fe^{3+} + 2H_2O \qquad (45)$$

Ferritin converts Fe^{2+} to less reactive Fe^{3+} from H_2O_2 and O_2. Tf ferritin is an important antioxidant, considering that the hydroxyl $\cdot OH$ radical is generated by the Fenton reaction from Fe^{2+} and H_2O_2. Considering that the components of ETC are greatly reduced when oxygen is needed as an electron acceptor O_2, the mobilization of iron from ferritin is also the result of an increase in the intracellular redox potential.

2.2.1.3. Heme Oxygenase

The main enzyme involved in *heme* group recycling is hemeoxygenase -HO which is controlled by several mechanisms. By cleavage of the *heme* ring at the α-methane bond to form biliverdin, it catalyzes the first step and rate-limiting cleavage of the *heme* group.

$$heme + NAD(P)H + H^+ + 3O_2 \leftrightarrow biliverdin + Fe^{2+} + CO + NAD(P)^+ + 3H_2O \quad (46)$$

NAD(P)H maintains the *heme* substrate of the central iron in a reduced state. The human HO protein is composed of 228 amino acids and is significantly inducible by heavy metals such as bromobenzene, as well as by hypoxia, endotoxins, cytokinesis, UV radiation, oxidative stress, and its *heme* substrate. It exists as a homodimer in many microorganisms and is found in the endoplasmic reticulum of eukaryotic cells.

Alcohol dehydrogenase catalyzes the ethanol oxidation to generate a larger amount of the reducing agent NADH, which is a cofactor in the production of iron-dependent ·OH hydroxyl radicals. From ferritin, NADH facilitates the mobilization of iron and ethanol is activated to the hydroxyethyl radical whose formation may be iron dependent. During ethanol metabolism, the iron-dependent acetaldehydoxanthine oxidase system initiates lipid peroxidation [19].

2.2.1.4. Hemoglobin and Myoglobin

The polypeptide chains of hemoglobin and myoglobin can autooxidize just like the iron in their *heme* groups. When hemoglobin iron and myoglobin bind O_2, it is normally as ferrous iron, but some electronic delocalization that exists allows this balance (47) to be achieved.

$$heme\text{-}Fe^{2+}\text{-}O_2 \rightleftharpoons heme\text{-}Fe^{3+}\text{-}O^{·-}_2 \tag{47}$$

These oxygenated molecules sometimes decompose to produce superoxide radical and methemoglobin or metamioglobin (*heme*-Fe^{3+}), which cannot bind to oxygen. Normally, only 3% of the hemoglobin in red blood cells is as methemoglobin. The oxidation of hemoglobin and myoglobin can be accelerated by transition metals or by nitrite (NO_2^-). In rural zones where nitrates (NO_3^-) are over-fertilized, they can pass into the intestine and be reduced by intestinal bacteria to nitrites, which can be absorbed and cause sufficient methemoglobin to interfere with body tissues oxygenation. The abnormal hemoglobins present higher percentages of methemoglobin.

2.3. COPPER

By electron spin resonance spectroscopy -ERS, the redox properties of the copper-iron complexes of bleomycin, adriamycin and thio-semicarbazones have been studied. These compounds may be readily reduced by thiol compounds and oxidized by iron or reduced iron species to form radicals. Some enzymes such as oxygenases and oxidases have copper as a cofactor. Copper catalyzes reactive oxygen species (ROS) formation and membrane lipid peroxidation in the same way as iron.

$$Cu(II) + O^{-}_{2} \rightarrow Cu(II) + O_{2} \tag{48}$$

$$Cu(I) + H_{2}O_{2} \rightarrow Cu(II) + {\cdot}OH + OH^{-} \tag{49}$$

$$O^{-}_{2} + 2H^{+} \rightarrow H_{2}O_{2} + O_{2} \tag{50}$$

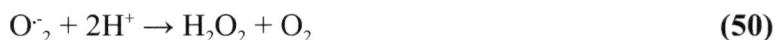

In a concentration-dependent manner, copper accelerates the oxidation of hydroquinone to benzoquinone. Oxidants and/or chelating agents such as glutathione and dithiothrethiol prevent cytotoxicity of hydroquinone by copper. The ability of glutathione to stabilize copper Cu(I), which prevents redox cycling and radical generation is one of the protective effects of glutathione.

2.4. CHROME

Chromium occurs mainly in the valence form Cr(VI) and Cr(III). The chromate ion $[CrO_{4}]^{-2}$, which is the dominant form of Cr(VI) under neutral aqueous conditions, through non-specific anionic transporters can cross the membrane very rapidly.

In the reduction of Cr(VI) by glutathione reductase, in the presence of NADPH and with the generation of hydroxyl radicals, the formation of Cr(V) intermediates has been demonstrated by the ESR technique. The Cr(V) complex produced reacts with hydrogen peroxide to generate hydroxyl radicals, which can be initiators of Cr(VI) cytotoxicity (Reactions 40, 41, 42, 51 and 52).

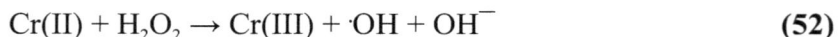

$$Cr(III) + O^{-}_{2} \rightarrow Cr(II) + O_{2} \tag{51}$$

$$Cr(II) + H_{2}O_{2} \rightarrow Cr(III) + {\cdot}OH + OH^{-} \tag{52}$$

In the presence of biological reductants such as NADH and L-cysteine, Cr(III) is reduced to Cr(II), which in turn reacts with $H_{2}O_{2}$ to generate ${\cdot}OH$ radicals, as detected by ESR and HPLC. Similarly, at physiological pH the incubation of Cr(III) with hydrogen peroxide generates hydroxyl radicals. Diethylenetriamine pentaacetic acid -DTPA considerably reduces hydroxyl radicals, while NADH, glutathione and L-cysteine have minimal effects. When Cr(III) is incubated with *tert*-butyl hydroperoxide and cumene hydroperoxide, lipid peroxide-derived free radicals are generated.

Although Cr(VI) is more toxic and produces considerable oxidative stress, the biologically active oxidation states are Cr(VI) and Cr(III). Both states participate in the redox cycle with the production of ROS [21].

2.5. CADMIUM

Cadmium is a non-essential element that is abundant in the environment due to industrial practices. The accumulation of cadmium salts generates toxicity in tissues, mainly due to lipid peroxidation. High concentrations of cadmium acetate inhibit SOD activity, increase lipid peroxide levels, and cause a decrease in glutathione levels due to the generation of ROS at a faster rate than the reduced glutathione regeneration.

Cd^{2+}-dependent ROS production at the plasma membrane and in the plant mitochondrial electron transport chain *in vitro* and *in vivo* was evaluated by Heyno *et al*, 2008 [22]. They observed that Cd^{2+} in the plasma membrane inhibited superoxide production, while in the mitochondria it increased superoxide and hydrogen peroxide production.

The thiol -SH groups of protein cysteines are the main targets of cadmium, so that inactivation of these groups in enzymes leads to deficiencies in the nucleus, mitochondria, and endoplasmic reticulum. Cadmium disintegrates the mitochondrial membrane potential and causes inhibition of cellular respiration because in ETC-mitochondrial cadmium inhibits electron flow through complex III cytochrome bc_1 or ubiquinone: cytochrome c reductase [23].

2.6. VANADIUM

There is a strong relationship between vanadium-induced hepatotoxicity and the induction of lipid peroxidation. In microsomes the one-electron reduction of V(V) takes place in the presence of NADH with the reduction of molecular oxygen to H_2O_2. This in turn reacts with V(V) to generate ·OH hydroxyl radicals. The ·OH hydroxyl radicals are generated by a Fenton-type reaction.

$$V(V) + O^{·-}_2 \rightarrow V(IV) + O_2 \ (53) \qquad\qquad \textbf{(53)}$$

$$V(IV) + H_2O_2 \rightarrow V(V) + ·OH + OH^- \ (54) \qquad\qquad \textbf{(54)}$$

When a mixture of V(V), glutathione reductase, ferredoxin-NADP$^+$ oxidoreductase, NADPH and lipoyl dehydrogenase is incubated in a phosphate buffer, V(V) species accumulate. Thus, the flavoenzymes mentioned above act as NADPH-dependent vanadium V(V) reductase.

The predominant oxidation state of vanadium inside the cell is tetravalent vanadyl generated by cellular reductants such as NADH, ascorbic acid, or glutathione by pentavalent vanadium conversion to vanadyl. Whereas, the vanadium oxidation to vanadate is carried out by oxidants such as NAD^+, O_2^{-2} and O_2 [24].

2.7. MERCURY

Mercury causes a decrease in the activities of enzymes responsible for the protection of the cell against the peroxidative action of hydroperoxides and superoxide anions. Such enzymes are glutathione, glutathione peroxidase, catalase, and SOD.

Considering the above, then, according to mercury concentrations and incubation time, mitochondrial glutathione reductase levels decrease. Thus, at low mercury concentrations 12-30 nmol/mg protein, mitochondrial glutathione decreases and increases the formation of H_2O_2 in the mitochondrial ETC, this time damaged.

It has been demonstrated that methylmercury and ethylmercury are rapidly degraded by the xanthine oxidase-hypoxanthine, EDTA-iron, and copper ascorbate systems. It was found that ·OH hydroxyl radicals may be responsible for the removal of the two forms of organic mercury and that this removal is not related to the production of superoxide or hydrogen peroxide.

S-Hg-R complexes are formed by mercury reaction with active sulfhydryl -SH groups. The toxicological behavior of mercury is due to the affinity of MeHg for the anionic form of the -SH groups, where log K is extremely high on the order of 15-23. Unlike the affinity of this compound for ligands containing oxygen, nitrogen, or chlorine, as well as for amino or carboxyl groups where the constant is 10 orders of magnitude lower [25].

Numerous investigations have been carried out on the mechanism of oxidation of environmental Hg^0, by O_3 and OH, informing on the associated kinetics and its speed constants, determined under suitable atmospheric conditions. Recent works have incorporated these mechanisms in the validation of models, which has made it possible to know the function of O_3/OH oxidation in the atmosphere [26].

2.8. NICKEL

In the atmosphere, nickel is mainly bound to oxygen and sulfur to form oxides or sulfides in the earth's crust. The excessive production of nickel at an industrial level has transformed it into one of the highly polluting heavy metals of the earth's atmosphere. Nickel, like most heavy metals, produces free radicals such as superoxide anions, generating acute and chronic toxicities in living beings [27].

Alkyl, alkoxyl and peroxyl radical formation occurs when nickel reacts with *tert*-butyl hydroperoxide or cumene hydroperoxide in the presence of antioxidants such as glutathione, anserine, homocarnosine and carnosine. These oligopeptides may facilitate nickel-mediated free radical production, rather than having a protective effect against oxidative damage.

2.9. ZINC

Zinc acts as a membrane stabilizer and prevents the formation of reactive oxygen species - ROS, unlike some cations that facilitate the formation of oxidative stress. This is done by a mechanism of displacement of metal ions from the specific redox site, which implies a protection of sulfhydryl groups against oxidation and inhibition by metal ions of ROS production.

2.10. LEAD

The lead atom, by its electronic properties, can form covalent bonds with groups such as sulfhydryl and antioxidant enzymes, usually inactivating them. For example, it inactivates GSH, catalase and SOD.

Superoxide radical scavenging is decreased when SOD is reduced, whereas the ability to trap superoxide is affected when CAT is reduced. Lead replaces zinc ions as important cofactors for these enzymes. In addition to the above, lead in its ionic form substitutes monovalent cations such as Na^+ and divalent cations such as Ca^{2+}, Fe^{2+} and Mg^{2+}, thus affecting the function of such ions in the vital processes of living organisms [28].

2.11. COBALT

Cobalt ions have been demonstrated to inhibit the *heme* protein of the electron transport chain by direct interaction of the cobalt ions with the iron atom of the protein. It also inhibits the processing of the precursor unit of cytochrome *c* oxidase COX in the mitochondria, degrading it. Cobalt also binds especially to Ca^{2+}-binding proteins and displaces other ions such as Mg^{2+}, Fe and Zn.

Cell viability and superoxide formation in the cytoplasm and mitochondria after incubation of different cell types with cobalt ions were evaluated by Chameon *et al.* 2019 [29]. They found that in blood cells $CoCl_2$ generates free radical formation and in monocytic cells it forms mitochondrial superoxide and cytoplasmic superoxide.

2.12. XENOBIOTICS. PARAQUAT – PQ

Paraquat -PQ (1,1'-4,4'-bipyridium dichloride; methyl viologen), is one of the most widely applied non-selective bipyridyl herbicides. It is a contact herbicide with a broad spectrum of action, toxic to humans and animals and, in some cases, a potent poison.

The molecular mechanisms of PQ toxicity are still unclear; however, two mechanisms may be involved in its activity: PQ-induced mitochondrial toxicity and PQ-related cellular toxicity. The latter is due to the production of reactive oxygen species-ROS generated from PQ metabolism by the microsomal enzyme system.

Considering that mitochondria are the main source of reactive oxygen species-ROS in cells, inhibition of ETC increases the levels of these autooxidizable components and thus increases the generation of reactive oxygen species-ROS from mitochondria.

Paraquat -PQ can accept electrons from complex (I). The PQ radical formation has been observed *in vitro*. The site of PQ radical formation is around the 30 kD subunit of complex (I). This subunit is a transmembrane protein. However, because PQ is very soluble in water, it has difficulty in crossing the mitochondrial inner membrane. The reaction between NADH and PQ is catalyzed by complex (I) utilizing NADH in the matrix and PQ in the intermembrane space between the mitochondrial inner and outer membranes.

Through oxidation and reduction processes, PQ can participate in the mitochondrial redox system. The main oxidation states of PQ are:

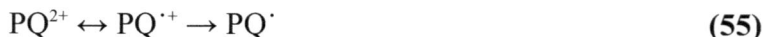

$$PQ^{2+} \leftrightarrow PQ^{\cdot+} \rightarrow PQ^{\cdot} \tag{55}$$

A colorless divalent cation is PQ. In turn, the partially reduced $PQ^{\cdot+}$ contains an unpaired electron that is shared by all the central carbon positions of the ring. The reaction is reversible, so that an equivalent of the reducing agent reduces more than 50% of $PQ^{\cdot+}$ to $PQ^{\cdot 2+}$ if its reduction potential is more negative than that of PQ. Therefore, PQ^{2+} potential for the first reduction is -0.446V and -0.88V for the second reduction.

The divalent cation PQ^{2+} accepts electrons in aqueous solution to form the blue radical $PQ^{\cdot+}$, which rapidly reacts with molecular oxygen to generate O^{-}_{2} superoxide radical (Fig. **19**). For its part, $PQ^{\cdot+}$ promotes the continuous formation of superoxide by presenting repeated redox cycles, under aerobic conditions.

Paraquat -PQ derivatives extract electrons from the intermediates of the respiratory chain, in a similar way as natural substrates perform. They also affect the redox cycle in the last step, returning electrons to the respiratory chain and thus generate energy [30].

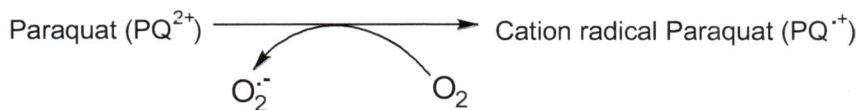

Fig. (19). Redox cycle of the PQ radical.

The oxidation-reduction cycle of PQ initiates lipid peroxidation through superoxide radical and singlet oxygen intermediates, thus constituting the second proposed mechanism for PQ mitochondrial toxicity. Therefore, the main factors of mitochondrial damage caused by PQ involve: the cascade of chemical reactions leading to PQ reduction, the generation of free radicals and lipid peroxidation (Fig. **20**).

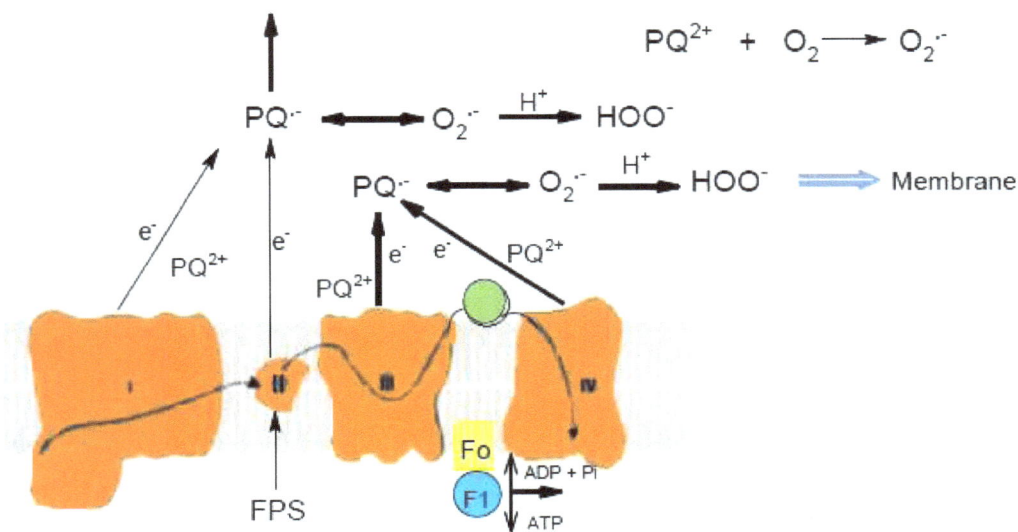

Fig. (20). PQ toxicity mechanism.

Mitochondrial processes and the amount of hMnSOD protein precursors are events susceptible to PQ-induced dysfunction. In addition, PQ inhibits human MnSOD-dependent processes (section 1.1.1. Chapter 4). On the other hand, MnSOD-deficient cells are less sensitive to oxygen toxicity than CuZnSOD-deficient cells (Fig. **21**). PQ toxicity has been attributed to H_2O_2 production because membranes are considerably permeable to H_2O_2. Catalase protects against PQ toxicity, however, PQ generates H_2O_2 and also decreases catalase activity.

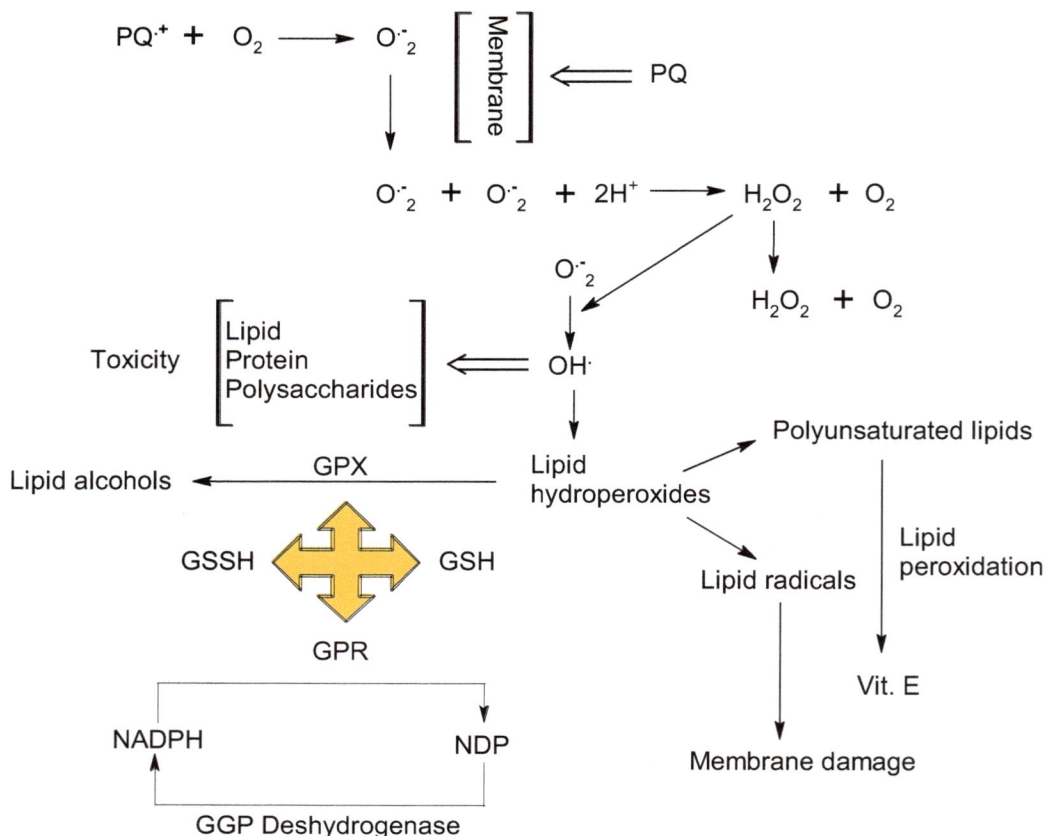

Fig. (21). Enzymatic and non-enzymatic metabolism of PQ in mitochondria.

The GSH is the most important mitochondrial antioxidant and its reduction may enhance the sensitivity of the mitochondrial structure to ROS-mediated damage. Among oxidative stress agents, PQ is a thiol oxidizing agent resulting from the rapid oxidation of reduced glutathione (GSH) to oxidized glutathione (GSSG). PQ induces lipid peroxidation, inhibits the mitochondrial redox chain and ATP synthase activity, and thus uncouples oxidative phosphorylation.

CONCLUSION

Analysis of reaction kinetics and one- and two-electron transfer mechanisms as well as electron and proton transfer mechanisms involved in the enzymatic oxidation processes of various systems that are sources of reactive oxygen species

(ROS) was performed. Reactive oxygen species are also produced by various organelles, cytosolic molecules, and exogenous sources such as heavy metals and xenobiotics.

REFERENCES

[1]   Martínez, C.M. Estrés oxidativo y mecanismo de defensa antioxidante.En: GIL H. A. In: *Tratado de nutrición*; Tomo I. Editorial Panamericana. 2010; pp 455-480.

[2]   Ohnishi, T.; Yamazaki, H.; Iyanagi, T.; Nakamura, T.; Yamazaki, I. One-electron-transfer reactions in biochemical systems. II. The reaction of free radicals formed in the enzymic oxidation. *Biochim. Biophys. Acta,* **1969**. *172*(3), 357-369.
[http://dx.doi.org/10.1016/0005-2728(69)90132-7] [PMID: 4305692]

[3]   Sidahmed-Adrar, N.; Marchetti, C.; Bonnefont-Rousselot, D.; Thariat, J.; Onidas, D.; Jore, D.; Gardes-Albert, M.; Collin, F. Interaction between non-anionic phospholipids and cytochrome c induced by reactive oxygen species. *Chem. Phys. Lipids,* **2010**, *163*(6), 538-544.
[http://dx.doi.org/10.1016/j.chemphyslip.2010.04.002] [PMID: 20398641]

[4]   Doonan, C.J.; Stockert, A.; Hille, R.; George, G.N. Nature of the catalytically labile oxygen at the active site of xanthine oxidase. *J. Am. Chem. Soc.,* **2005**, *127*(12), 4518-4522.
[http://dx.doi.org/10.1021/ja042500o] [PMID: 15783235]

[5]   Olson, J.S.; Ballou, D.P.; Palmer, G.; Massey, V. The mechanism of action of xanthine oxidase. *J. Biol. Chem.,* **1974**, *249*(14), 4363-4382.
[http://dx.doi.org/10.1016/S0021-9258(19)42428-9] [PMID: 4367215]

[6]   Barber, M.J.; Siegel, L.M. Oxidation-reduction potentials of molybdenum, flavin, and iron-sulfur centers in milk xanthine oxidase: variation with pH. *Biochemistry,* **1982**, *21*(7), 1638-1647.
[http://dx.doi.org/10.1021/bi00536a026] [PMID: 6896281]

[7]   Nakamura, S.; Yamazaki, I. One-electron transfer reactions in biochemical systems. IV. A mixed mechanism in the reaction of milk xanthine oxidase with electron acceptors. *Biochim. Biophys. Acta,* **1969**, *189*(1), 29-37.
[http://dx.doi.org/10.1016/0005-2728(69)90221-7] [PMID: 4309792]

[8]   Van Heuvelen, A. Electron transport in xanthine oxidase. A model for other biological electron transport chains. *Biophys. J.,* **1976**, *16*(8), 939-951.
[http://dx.doi.org/10.1016/S0006-3495(76)85744-X] [PMID: 938732]

[9]   Harrison, R. Structure and function of xanthine oxidoreductase: where are we now? *Free Radic. Biol. Med.,* **2002**, *33*(6), 774-797.
[http://dx.doi.org/10.1016/S0891-5849(02)00956-5] [PMID: 12208366]

[10]   Whittaker, J.W. The radical chemistry of galactose oxidase. *Arch. Biochem. Biophys.,* **2005**, *433*(1), 227-239.
[http://dx.doi.org/10.1016/j.abb.2004.08.034] [PMID: 15581579]

[11]   Hawkins, C.L.; Rees, M.D.; Davies, M.J. Superoxide radicals can act synergistically with hypochlorite to induce damage to proteins. *FEBS Lett.,* **2002**, *510*(1-2), 41-44.
[http://dx.doi.org/10.1016/S0014-5793(01)03226-4] [PMID: 11755528]

[12]   Bylund, J.; Brown, K.L.; Movitz, C.; Dahlgren, C.; Karlsson, A. Intracellular generation of superoxide by the phagocyte NADPH oxidase: how, where, and what for? *Free Radic. Biol. Med.,* **2010**, *49*(12), 1834-1845.
[http://dx.doi.org/10.1016/j.freeradbiomed.2010.09.016] [PMID: 20870019]

[13]   Echtay, K.S. Mitochondrial uncoupling proteins--what is their physiological role? *Free Radic. Biol. Med.,* **2007**, *43*(10), 1351-1371.
[http://dx.doi.org/10.1016/j.freeradbiomed.2007.08.011] [PMID: 17936181]

[14]   Velayutham, M.; Hemann, C.; Zweier, J.L. Removal of H_2O_2 and generation of superoxide radical: role of cytochrome c and NADH. *Free Radic. Biol. Med.,* **2011**, *51*(1), 160-170.
[http://dx.doi.org/10.1016/j.freeradbiomed.2011.04.007] [PMID: 21545835]

[15]   Koolman, J.; Roehm, K.H. Color atlas of biochemistry. Thieme, **2005**; p. 467.

[16]   Iyanagi, T.; Yamazaki, I. One-electron-transfer reactions in biochemical systems. 3. One-electron reduction of quinones by microsomal flavin enzymes. *Biochim. Biophys. Acta,* **1969**, *172*(3), 370-381.
[http://dx.doi.org/10.1016/0005-2728(69)90133-9] [PMID: 4388705]

[17]   Iyanagi, T.; Yamazaki, I. One-electron-transfer reactions in biochemical systems. V. Difference in the mechanism of quinone reduction by the NADH dehydrogenase and the NAD(P)H dehydrogenase (DT-diaphorase). *Biochim. Biophys. Acta,* **1970**, *216*(2), 282-294.
[http://dx.doi.org/10.1016/0005-2728(70)90220-3] [PMID: 4396182]

[18]   Leonard, S.S.; Harris, G.K.; Shi, X. Metal-induced oxidative stress and signal transduction. *Free Radic. Biol. Med.,* **2004**, *37*(12), 1921-1942.
[http://dx.doi.org/10.1016/j.freeradbiomed.2004.09.010] [PMID: 15544913]

[19]   Chepelev, N.L.; Willmore, W.G. Regulation of iron pathways in response to hypoxia. *Free Radic. Biol. Med.,* **2011**, *50*(6), 645-666.
[http://dx.doi.org/10.1016/j.freeradbiomed.2010.12.023] [PMID: 21185934]

[20]   De Freitas, J.M.; Meneghini, R. Iron and its sensitive balance in the cell. *Mutat. Res.,* **2001**, *475*(1-2), 153-159.
[http://dx.doi.org/10.1016/S0027-5107(01)00066-5] [PMID: 11295160]

[21]   Stohs, S.J.; Bagchi, D. Oxidative mechanisms in the toxicity of metal ions. *Free Radic. Biol. Med.,* **1995**, *18*(2), 321-336.
[http://dx.doi.org/10.1016/0891-5849(94)00159-H] [PMID: 7744317]

[22]   Heyno, E.; Klose, C.; Krieger-Liszkay, A. Origin of cadmium-induced reactive oxygen species production: mitochondrial electron transfer *versus* plasma membrane NADPH oxidase. *New Phytol.,* **2008**, *179*(3), 687-699.
[http://dx.doi.org/10.1111/j.1469-8137.2008.02512.x] [PMID: 18537884]

[23]   Genchi, G.; Sinicropi, M.S.; Lauria, G.; Carocci, A.; Catalano, A. The effects of cadmium toxicity. *Int. J. Environ. res. public health,* **2020**, *17*(11), 3782.
[http://dx.doi.org/10.3390/ijerph17113782] [PMID: 32466586]

[24]   Zwolak. I. protective effects of dietary antioxidants against vanadium-induced toxicity: A review. *Oxid. Med. Cell. Longev.,* **2020**, 1490316.

[25]   Aschner, M.; Syversen, T. Methylmercury: recent advances in the understanding of its neurotoxicity. *Ther. Drug Monit.,* **2005**, *27*(3), 278-283.
[http://dx.doi.org/10.1097/01.ftd.0000160275.85450.32] [PMID: 15905795]

[26]   Lyman, S.N.; Cheng, I.; Gratz, L.E.; Weiss-Penzias, P.; Zhang, L. An updated review of atmospheric mercury. *Sci. Total Environ.,* **2020**, *707*, 135575.
[http://dx.doi.org/10.1016/j.scitotenv.2019.135575] [PMID: 31784172]

[27]   Das, K.K.; Reddy, R.C.; Bagoji, I.B.; Das, S.; Bagali, S.; Mullur, L.; Khodnapur, J.P.; Biradar, M.S. Primary concept of nickel toxicity - an overview. *J. Basic Clin. Physiol. Pharmacol.,* **2018**, *30*(2), 141-152.
[http://dx.doi.org/10.1515/jbcpp-2017-0171] [PMID: 30179849]

[28]   Assi, M.A.; Hezmee, M.N.M.; Haron, A.W.; Sabri, M.Y.; Rajion, M.A. The detrimental effects of lead on human and animal health. *Vet. World,* **2016**, *9*(6), 660-671.
[http://dx.doi.org/10.14202/vetworld.2016.660-671] [PMID: 27397992]

[29]   Chamaon, K.; Schönfeld, P.; Awiszus, F.; Bertrand, J.; Lohmann, C.H. Ionic cobalt but not metal particles induces ROS generation in immune cells *in vitro*. *J. Biomed. Mater. Res. B Appl. Biomater.,*

2019, *107*(4), 1246-1253.
[http://dx.doi.org/10.1002/jbm.b.34217] [PMID: 30261124]

[30]　Mohammadi-Bardbori, A.; Ghazi-Khansari, M. Alternative electron acceptors: Proposed mechanism of paraquat mitochondrial toxicity. *Environ. Toxicol. Pharmacol.,* **2008**, *26*(1), 1-5.
[http://dx.doi.org/10.1016/j.etap.2008.02.009] [PMID: 21783880]

CHAPTER 4

Antioxidant Defense Systems

Abstract: The enzymatic and non-enzymatic reaction mechanisms of primary and secondary defense systems developed by cells to diminish the effects caused by overproduction of reactive oxygen species-ROS as a metabolic response to the damaging effects from endogenous and environmental factors are presented here. Enzymatic reaction mechanisms developed by plants as an antioxidant defense system are also presented.

Keywords: Antioxidants, Catalase, Coenzyme Q-CoQ, Cu-ZnSOD, Glutathione, HNE, Plant antioxidant defense, Superoxide dismutase-SOD, Trapping free radicals, UCPs, Vitamin C, Vitamin E.

INTRODUCTION

To counteract the effect of oxygen-free radicals, the biological systems contain molecules of both enzymatic and non-enzymatic nature that constitute the so-called *antioxidant defense system*. These antioxidant defense systems function very efficiently in a coordinated manner and is responsible for the maintenance of cellular homeostasis against oxidative stress generated from reactive oxygen species-ROS and other radicals originated during oxygen metabolism.

A distinction is made between primary or preventive antioxidant defense systems and secondary or chain-breaking antioxidant defense systems. The primary defense system directly reacts with the reactive oxygen species -ROS and thus reduce the initiation rate of free radical reactivations. The secondary defenses trap the propagating radicals, stopping their deleterious effect in the initial stages.

1. PRIMARY ANTIOXIDANT DEFENSE SYSTEMS

1.1. Enzymes

There are several enzymes whose primary function is to decrease intracellular and intercellular concentrations of reactive oxygen species. These include superoxide dismutases, catalase, glutathione peroxidase, glutathione reductase, glucose

-6-phosphate dehydrogenase, among others. Fig. (**1**) summarizes the concerted action of some of these intracellular enzymes.

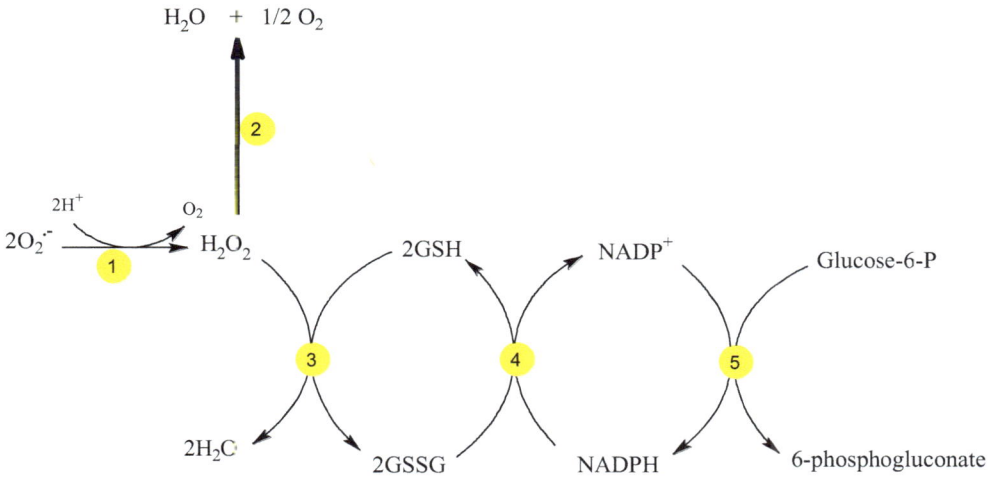

Fig. (1). Antioxidant Defense Enzyme Systems. **1.** SOD Superoxide dismutase, **2.** Catalase, **3.** GSH peroxidase, **4.** GSH reductase, **5.** Glucose 6-P-dehydrogenase.

1.1.1. Superoxide Dismutase-SOD

Superoxide dismutase (SOD) is a metalloenzymes family, which are found in almost all oxygen-exposed organisms and catalyzes the $O^{.-}_2$ dismutation reaction to produce H_2O_2 and O_2 (reaction 1). The spontaneous dismutation rate at physiological pH is much lower, about 10^4 times, than the dismutation rate of the reaction catalyzed by SOD. The SOD catalytic action was discovered by McCord and Fridovich. The classification of SOD depends on the great variety of prosthetic groups they possess.

$$O^{.-}_2 + O^{.-}_2 + 2H^+ \rightarrow H_2O_2 + O_2 \qquad (1)$$

Enzymes such as catalases and peroxidases transform H_2O_2 into a stable aqueous product.

The existence of SOD is common to all life forms. Their production and evolution as an antioxidant enzyme are linked to O_2 production by photosynthetic organisms approximately 2.000 Ma ago, through oxygen metabolism. In the SOD family, different metal centers have been detected, mainly Cu, Zn-, Fe-, Mn- and Ni-SODs. However, in prokaryotes two forms Cu,Zn SODs and Fe SODs/Mn SODs appear independently.

In chloroplasts, eukaryotes and bacteria mainly Cu,Zn SOD is found. Two different forms of Cu,ZnSOD such as cytosolic SOD1 and extracellular SOD/SOD3 are found in animals, where SOD1 differs from SOD3 in terms of amino acid composition and molecular weight. In turn, MnSOD is found in mitochondria of eukaryotes and prokaryotes. In addition to chloroplasts and mitochondria SOD accumulates in cytosol, extracellular matrix, glyoxysomes, peroxisomes and microsomes or any compartment where O_2 can be activated. Since O_2^- cannot cross the membrane, it can be eliminated at the site where it is produced.

The following reaction sequence describes the SOD catalytic mechanism:

$$M^{3+} + O_2^- + H^+ \rightarrow M^{2+}(H^+) + O_2 \tag{2}$$

$$M^{2+}(H^+) + H^+ + O_2^- \rightarrow M^{3+} + H_2O_2 \tag{3}$$

Where M, represents a metallic cofactor.

The stepwise sequence in which the SOD reaction proceeds presents several thermodynamic advantages: 1) With only one molecule at a time, the reactant overcomes the repulsion of the electrostatic potential between the two O_2^- anions. 2) The positively charged metals mediate specific binding with the negatively charged O_2^- at the active site. 3) The capture of a proton in the second step, through the reduction of metal ions, preserves the electrostatic attraction of the active sites. Considering the neutrality of the disproportionation products and that by this mechanism they do not bind and 4) The energy released in the first (thermodynamically favorable) half-reaction is utilized for O_2^- reduction of the second step [1].

SOD presents interesting properties such as: 1) The reaction rate ($M^{-1}s^{-1}$) of Mn-SOD is 6.8 x 10^8 and 6.6 x 10^8 for Fe-SOD in *Escherichia coli*. respectively and for erythrocyte Cu,ZnSOD is 6.4 x 10^9. The electron transfer between the substrate and the active site usually reaches the desired value and the catalytic rate reaches the diffusion limit. 2) SODs present high stability to factors such as free thawing, high temperatures, urea, and unfavorable pH.

Cu,Zn SODs are homodimeric, each monomer containing one atom of Cu and Zn and a molecular weight of 14-33 KDa. Their catalytic activity is resistant to cleavage by the enzyme proteinase K; to physical treatments such as heat, thawing and cold cycles; and to chemical treatments such as 4% SDS and 8M urea. They are inhibited by H_2O_2, amide, diethyldithiocarbamate and cyanide. Due to their homotetrameric nature, Cu,Zn SODs are different between humans and other mammals. Extracellular SODs have some differences from cytoplasmic SODs:

1) although the amino acid residues involved in the coordination of Cu and Zn are conserved, the central region of extracellular SOD is approximately 50% identical to the last two thirds of SOD1, 2) extracellular SOD does not react with antibodies produced against SOD1 and 3) the extracellular SOD localization is due to a strong affinity for heparin present on cell surfaces.

In a manner similar to Cu,ZnSODs, the Fe-SODs and Mn-SODs are homodimers or tetramers containing one metal atom per subunit of 13-44 kDa. While Cu,ZnSODs are dispensable, Mn-SOD is essential in biological processes since its main function is to trap O^{-}_{2} to preserve mitochondrial membrane integrity. However, due to its structure containing nine *Tyr* residues, some of which are oxidizable, MnSOD is very sensitive to oxidative inactivation. Whereas human Cu,ZnSOD does not contain *Tyr* residues and is more resistant to ROS. Thus, MnSOD alterations can occur very rapidly under pathological conditions.

ROS in plants are important signaling molecules in response to biotic and abiotic factors. Their production is also activated as a defense mechanism in the presence of some symbiont fungi such as Vesicular-Arbuscular Mycorrhizal fungi (VAM). The SOD function in this case is to protect the symbiotic system since ROS can damage both the symbiont fungus and the host plant cell. Plants can also be protected from high altitude cold stress and ozone damage by increased SOD levels. Likewise, plants can be tolerant to salinity because SOD produces lignification of vascular structures, which in turn is also induced by H_2O_2. Slow root growth, which affects mitochondrial redox homeostasis and tricarboxylic acid flux, occurs when MnSOD is suppressed.

In marine algae, ROS are also signaling molecules that allow to detect self-induction, reduce competition for bacterial infection or iron availability. However, algae protect themselves from the damaging effects from photochemical ROS caused by light intensities through the production of high SOD contents. On the other hand, *Synechococcus* cyanobacteria, by means of Fe-SOD, protect themselves from high levels of ROS caused by cold stress.

In addition to the protective effect of SOD against oxidative stress caused by ROS in higher organisms, SOD is also present in microbes, for example in O_2-metabolizing aerobic microorganisms but is absent in some strict anaerobes. In bacteria for example, Cu,ZnSOD protects periplasmic proteins from endogenous O^{-}_{2}. Enzyme production is induced in the aerobic stationary phase for survival. Pathogens also produce Cu,ZnSOD to protect themselves from host cell ROS. Therefore, periplasmic Cu,ZnSOD traps ROS produced in the oxidative burst of phagocytes. Thus, antibody blocking, overproduction of SOD and addition of exogenous SOD are some of the defense systems employed by bacterial and eukaryotic pathogens against host cells.

Xu and Kuppusamy 2005 [2] evaluated relationship between model enzyme dCu,ZnSOD and ·OH hydroxyl radical production employing spin trapping reagent DMPO. They demonstrated that dCu,ZnSOD induces the generation of ·OH hydroxyl radicals independently of superoxide. Thus, Cu,ZnSOD can have a dual effect depending on the enzyme conditions.

The main component of the Cu,ZnSOD is copper, it has the complete external orbital which allows it to gain or lose electrons more easily. In addition, Cu catalyzes free radical production reactions. Meanwhile, Zinc does not form free radicals and only exists in one valence. In experiments to evaluate the ·OH hydroxyl radical production by dCu, ZnSOD the chelating reagents NaCN and DDC were employed and it was observed that Cu reacts with the cyanide ion (CN-) to form the following complex:

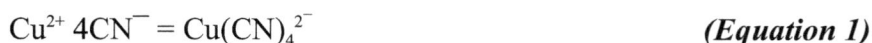

$$Cu^{2+} \, 4CN^- = Cu(CN)_4^{2-} \qquad \textit{(Equation 1)}$$

Cu also reacts with DDC to form the Cu(II)-DDC complex. The inhibition of ·OH production by dCu,ZnSOD by the reaction between Cu and NaCN or DDC indicates that O^-_2 is not involved. The involvement of H_2O_2 in the formation of ·OH induced by dCu,ZnSOD was demonstrated because the catalase enzyme that converts H_2O_2 to water and oxygen also scavenges the hydroxyl radical. Therefore, the participation of Cu and H_2O_2 in the increased activity of dCu,ZnSOD demonstrates that under different experimental conditions CuZnSOD releases Cu and general ·OH radicals.

SOD overproduction has been demonstrated to have deleterious effects, some of which are caused by the increased production of H_2O_2, which in turn could be related to the increased rate of O^-_2 dismutation. Thus, all O^-_2 could be converted to $H_2O_2 + O_2$ with the same neutral constant if the O^-_2 dismutation occurred by SOD catalysis with a constant of $\sim 10^9$ M^{-1} s^{-1} or if it were dismutated in a neutral aqueous medium with a constant of $\sim 10^5$ $M^{-1}s^{-1}$. Therefore, the SOD effect would be influenced by the O^-_2 concentration.

In contrast, H_2O_2 production decreases in the fraction of O^-_2 that reacts with cytochrome c to produce O_2 when the systems contain an O^-_2 oxidant such as ferric cytochrome c. In this case, SOD diverts O^-_2 into the dismutation reaction to increase H_2O_2 production when it competes with cytochrome c by O^-_2. However, the addition of SOD can decrease H_2O_2 production when the system contains superoxide reductase or an O^-_2 reductant such as the [4Fe-4S] group of aconitase. Considering that O^-_2 reduction generates one H_2O_2 for each O^-_2 consumed, whereas the dismutation reaction produces one-half, then the decrease in H_2O_2 is limited to a factor of 2.

Biological systems can produce O^{-}_2 in response to factors such as: the degree of reduction of complexes I and III of the mitochondrial electron transport chain, pO_2, the activities of O^{-}_2 producing enzymes such as xanthine oxidase, pyocyanin or quinones and the presence of exogenous compounds to the redox cycle such as paraquat, among others.

Numerous enzymes catalyze H_2O_2 production from the reduction of divalent O_2, for example xanthine oxidase under normal experimental conditions reduces ~80% of the O_2 consumed into H_2O_2, this enzyme being a source of O^{-}_2. This suggests that not all H_2O_2 generated by living systems comes from O^{-}_2.

Studies on the deleterious effects of higher-than-normal levels of SOD have been carried out to determine the effect of normal levels of this enzyme. Thus, it has been demonstrated that the effects caused by SOD excessive concentration are originated by: 1) Cu,ZnSOD acts as a reducing enzyme: superoxide oxidoreductase, as a non-specific peroxidase and as a cysteine oxidase, therefore SODs can perform activities not related to O^{-}_2 dismutation and for SOD hyperlevels, these activities can be deleterious, 2) Cu,ZnSOD overproduction limits the availability of Cu(II) needed for cytochrome oxidase formation therefore excessive metal-SOD production may compete with the biosynthesis of other metalloenzymes, 3) O_2 may react with the $L^{\cdot}$ lipid radical to produce the $LOO^{\cdot}$ peroxyl radical and O^{-}_2 plus a proton may also react with $L^{\cdot}$ to produce a LOOH, which is a relatively unreactive hydroperoxide, therefore O^{-}_2 levels exert a favorable effect on chain termination reactions [3].

As excessive SOD production causes deleterious effects by the peroxidase activity of Cu, ZnSOD or by the activity of SOD as superoxide reductase, which increases H_2O_2 formation, it could then be said that overproduction of SOD inevitably induces an increase in H_2O_2 production.

O^{-}_2 could control its own production by the following mechanism: Inactivation of aconitase by O^{-}_2 decreases NAD^+ reduction and reduces electron flow along the mitochondrial electron chain resulting in a decrease of the ground states in complex I and III reduction leading to a decrease in O^{-}_2 production. By decreasing the self-limitation of O^{-}_2 formation SOD could increase H_2O_2 formation. The peroxidation of some compounds is known to be catalyzed by Cu,ZnSOD.

The mechanism by which proton transport occurs during O^{-}_2 dismutation can be explained due to the lack of deuterium isotope effect: In the first interaction with O^{-}_2 a reduction of the Cu(II) site occurs and in the copper and zinc binding sites there is a breaking of the bond linking copper and imidazole residues. During the interval between the O^{-}_2 interactions, the Zn-bound imidazolate resulting from the first interaction may protonate. Then the solvent may enter the channel and reoxidize Cu(I) to SOD-Cu(II); thus, the bond can be restored by releasing the proton for the formation of outgoing HO_2^{-}. By X-ray crystallography, the release of the imidazole bond in the reduction of SOD-Cu(II) could be confirmed.

Regarding the effect of pH on SOD activity, Liochev and Fridovich 2010, have proposed that it is bicarbonate HCO_3^{-} that creates a binding site for H_2O_2 when it binds to the enzyme. Bicarbonate facilitates the oxidation of substrates to a diffusible carbonate radical $CO_3^{·-}$, including those substrates that enter the solvent access channel to the copper site. Thus, it has been demonstrated that in the presence of H_2O_2 and HCO_3^{-}, SOD produces $CO_3^{·-}$.

Through pulse radiolysis studies it was shown that $CO_3^{·-}$ is a strong oxidant that oxidizes substrates with constants between $\sim 10^5$ and 10^{10} $M^{-1}s^{-1}$. The ·OH hydroxyl radical can also oxidize HCO_3^{-} to produce the $CO_3^{·-}$ radical.

Between the effect of CO_2 and HCO_3^{-} on the increase of SOD peroxidase activity, CO_2 was found to have the greatest effect. The SOD addition means that H_2O_2 and HCO_3^{-} are prebalanced, while the H_2O_2 addition means that they have not been prebalanced, establishing that CO_2 is an essential reactant and is slowly removed in some form of CO_2 adduct plus H_2O_2, such as peroximonocarbonate (HCO_4^{-}).

Several mechanisms to produce radical carbonate have been proposed, involving the reduction of peroximonocarbonate by reduced SOD. The kinetic parameters have been calculated for H_2O_2 and HCO_3^{-}; however, it should be clarified that the real substrates involved in SOD peroxidase activity are HO_2^{-} and CO_2. Reaction scheme is as follows and the reaction constants involved are presented in Table **1** [4].

$$SOD - Cu(II) + HO_2^{-} \rightleftharpoons SOD - Cu(I) + H^{+} + O^{-}_2 \qquad (4)$$

$$SOD - Cu(II) + O^{-}_2 \rightleftharpoons Cu(I) + O_2 \qquad (5)$$

$$SOD - Cu(I) + HO_2^{-} + H^{+} \rightleftharpoons SOD - Cu(II)OH + HO^{-} \qquad (6)$$

$$SOD - Cu(II)OH + CO_2 \rightleftharpoons SOD - Cu(II) + CO_3^{·-} + H^{+} \qquad (7)$$

The oxidant SOD-Cu(II)OH bond, which can be written as SOD-Cu(I) or SOD-Cu(III), inactivates the enzyme because it reacts with the histidine residue in the copper ligand. It can also cause slow oxidation of various substrates, thus protecting SOD from inactivation. Considering that alcohols reacting with $\cdot$OH at a constant 10^9 $M^{-1}s^{-1}$ does not protect SOD against inactivation by H_2O_2, it is likely that the SOD-Cu(II)OH bond is different from the free $\cdot$OH. When CO_2 is present, this substrate is oxidized by the oxidative bond and produces $CO_3^{\cdot-}$, as is presented in reaction 7.

Table 1. Reaction constants.

No.	Reaction Constants	pH
4	50 $M^{-1}s^{-1}$	7.4
-4	2×10^9 $M^{-1}s^{-1}$	7.0 – 9.0
5	2×10^9 $M^{-1}s^{-1}$	7.0 – 9.0
-5	0.5 – 3.7 $M^{-1}s^{-1}$	7.4
6	13 $M^{-1}s^{-1}$	7.4
9	3.1×10^{-3} $M^{-1}s^{-1}$	7.2
-9	1×10^{-2} $M^{-1}s^{-1}$	7.2

A major part of $CO_3^{\cdot-}$ diffuses into free solution and may disappear by self-reaction, oxidation of amino acid residues exposed to SOD and oxidation of exogenous substrates, while a small portion of this $CO_3^{\cdot-}$ inactivates the SOD enzyme by oxidation of a histidine residue. Considering the above, CO_2 cannot protect SOD by inactivation by H_2O_2, since this inactivation is enhanced by high CO_2 contractions.

When CO_2 is present in reaction 7 due to the presence of the oxidizing bond, the equilibrium of the reaction tends to the right. CO_2 accelerates the complete substitution, causes a considerable increase in H_2O_2 consumption, and restores SOD-Cu (II). In the absence of CO_2 in reaction 6 the reaction equilibrium shifts to the left. However, at high values of $[CO_2]$, medium saturation at $\sim$ 100 mM HCO_3^-, and at pH 7.4, CO_2 moves the reaction equilibrium completely to the right, followed by k_5 by 13 $M^{-1}s^{-1}$ (Table **1**). This ability of CO_2 to increase SOD peroxidase is clear evidence against the formation of free OH in reaction 6. SOD is clearly a CO_2 peroxidase.

How SOD acts as an HCO_4^- reductase instead of a CO_2 peroxidase can be explained by mechanisms and how reactions 6 and 7 could be replaced by the following reaction 8

$$SOD - Cu(I) + HCO_4^- \rightleftharpoons SOD - Cu(II) + CO_3^{\cdot-} + HO^- \qquad (8)$$

In reaction 9, El HCO_4^- is formed spontaneously and in a reversible manner:

$$HCO_3^- + H_2O_2 \rightleftharpoons HCO_4^- + H_2O \tag{9}$$

The constants for reactions 9 and -9 are given in Table **1**. To determine the time to accumulate HCO_4^-, add SOD-Cu(I) and follow its oxidation to SOD-Cu(II), HCO_3^- and H_2O_2 were incubated. Performing a control without preincubation where H_2O_2 was added, it was demonstrated that it is H_2O_2 that oxidizes SOD-Cu(I), as confirmed by reactions 6 and 7. Then at pH 7.4 it was observed that in 5 to 6 minutes, one half of the HCO_4^- disappeared and after 30 to 40 minutes, it disappeared completely.

This slow consumption of HCO_4^- is due to the consumption of H_2O_2 by SOD, determined by reactions 4, -4, 5, 6 and 7, after a decrease of HCO_4^-, represented by reactions 9 and -9 according to equation 2 where the rate of exchange of $[HCO_4^-]$ is represented:

$$d[HCO_4^-]/dt = k_9[H_2O_2][HCO_3^-] - k_{-9}[HCO_4^-] \qquad \textbf{\textit{(Equation 2)}}$$

During peroxidase catalysis and SOD inactivation, the SOD-Cu(I)/SOD-Cu(II) ratio was slightly higher than 1, this possibly is the explanation for the observed time of HCO_4^- disappearance.

However, the observed time does not explain the possibility that SOD is a HCO_4^- reductase. In this case the reactions involved are: 4, -4, 5, 8, 9 and -9. Equation 3 represents the rate of HCO_4^- exchange.

$$d[HCO_4^-]/dt = k_9[H_2O_2][HCO_3^-] - k_{-9}[HCO_4^-] - k_8[SOD-Cu(I)][HCO_4^-] \qquad \textbf{\textit{(Equation 3)}}$$

The initially high amount of HCO_4^- could disappear ~ 30 times faster at high k_8, such as 150 $M^{-1}s^{-1}$. However, the rate of disappearance of HCO_4^- could be as reported here with low k_8. But if it disappears abruptly, the rate of H_2O_2 disappearance would be very slow, SOD could be inactivated and therefore the decrease in HCO_4^- would be insignificant.

Assuming that k_8 is equal to 150 $M^{-1}s^{-1}$ and little involvement of additional reactions, then SOD-Cu(I)/SOD-Cu(II) ratio could be approximately 1.

Bonini *et al.*, 2009 [5], concluded that SOD might reduce HCO_4^- and propose the following model where:

1. Ignores k_9 and k_{-9} of the processes that generate the disappearance of HCO_4^-

according to equation 4:

$$- d\,[HCO_4^-]/dt = k_8[SOD][HCO_4^-] \qquad\qquad \textbf{\textit{(Equation 4)}}$$

This causes SOD to react experimentally only with the pre-existing HCO_4^- so that it can disappear in seconds and not in minutes. In the real case from H_2O_2 and HCO_3^- the HCO_4^- can be produced and disappear within minutes if the SOD-Cu(I) is slightly more than one half of the [SOD] and the k_8 is 150 $M^{-1}s^{-1}$.

2. Furthermore to reaction 8, pathways are introduced for the disappearance of SOD-Cu(I) as well as reaction 10.

a. Besides reaction 8, disappearance of SOD-Cu(I) may occur by reaction 10:

$$SOD - Cu(I) + H_2O_2 \;\rightleftharpoons\; SOD - Cu(II) + HO^- + \cdot OH \qquad (10)$$

With k_{10} being 13 $M^{-1}s^{-1}$.

b. The k for reaction -4 is 1×10^{10} $M^{-1}s^{-1}$, while the k for reaction 5 is 2×10^9 $M^{-1}s^{-1}$. However, the value of the constant for reaction -4 is for the copper ion. The k for reactions -4 and 5 are found in Table 1 and are close to $2 \times 10^9 M^{-1}s^{-1}$.

c. In the SOD active site, $CO_3^{\cdot-}$ (product of reaction 8) reacts with reduced copper. This is a reaction that can explain the difficulties in detecting $CO_3^{\cdot-}$ by EPR in the presence of SOD plus HCO_3^- and H_2O_2. However, $CO_3^{\cdot-}$ can react with the chelators employed as well as with the amino acid residues of SOD.

d. HCO_4^- and $CO_3^{\cdot-}$ protect SOD from inactivation by limiting the availability of SOD-Cu(I), because SOD-Cu(I) disappears in another reaction leading to SOD inactivation. This protection may decrease when HCO_4^- is consumed, thus accelerating SOD inactivation.

3. The model predicts SOD inactivation in a given time. Initially the inactivation rate is rapid but then decreases with decreasing SOD concentration and H_2O_2 with or without HCO_3^-.

In experiments on arrested flow kinetics performed by Medina *et al.*, 2009 [6], and considering the constant dissociation of peroxymonocarbonate to H_2O_2 and HCO_3^- it was found that peroxymonocarbonate dissociates also when reacting with SOD-CU(II), and that it acts as an intermediate of the catalytic cycle. Comparing SOD peroxidase activity with substrates such as format, carbon dioxide bicarbonate or format plus carbon dioxide bicarbonate in 8.4 buffer demonstrated that format does not compete with the carbon dioxide bicarbonate pair when SOD is bound to the oxidant SODCu(I), SOD-Cu(III) or SOD-Cu(II)OH.

When carbon dioxide is added to the SOD-Cu(I) peroxide complex, peroxymonocarbonate is produced (reaction 11), which in turn is reduced to the radical carbonate and the enzyme returns to its initial state (reaction 12):

$$SOD - Cu(I) + HO_2^- \rightleftharpoons SOD - Cu(I)HOO^- \tag{11}$$

$$SOD - Cu(I)HOO^- + CO_2 \longrightarrow SOD - Cu(I)HCO_4^- \longrightarrow SODCu(II) + CO_3^{\cdot -} + HO^- \tag{12}$$

Dioxide is a carbonate radical producer due to the sensitivity of carbon dioxide to nucleophilic action by deprotonated/activated peroxides.

Considering the specificity of SOD by $O^{\cdot -}_2$, Liochev and Fridovich in 2010 determined the structure of SOD by X-ray crystallography and found that the solvent access to copper is a narrow channel together with an electrostatic field that attracts ions, such that access to copper from large molecules is difficult. Under neutral conditions with a constant of $\sim 2 \times 10^9$ $M^{-1}s^{-1}$, the $O^{\cdot -}_2$ reduces SOD-Cu(II), whereas with a constant of ~ 50 $M^{-1}s^{-1}$, the H_2O_2 reduces SOD-Cu(II). However, the influence form is HO_2^- and not H_2O_2 which has a *pKa* of ~11.6. The above means that with a constant of $10^6 M^{-1}s^{-1}$, the HO_2^- reduces the active site of Cu(II). Thus, with only the addition of one hydrogen atom to $O^{\cdot -}_2$, SOD-Cu(II) reduction rate is decreased approximately 2000-fold. Such a large difference may be due to the differences in the redox potential between $O^{\cdot -}_2$ and H_2O_2 and the increase in molecular weight from 32 to 33 Dalton. Considering that on the second step of the dismutation of $O^{\cdot -}_2$ by SOD the influent group is HO_2^-, then the *pKa* of H_2O_2 would be expected to be as high as 11.6.

CO_2 is the substrate for SOD because CO_2 imparts an induced polarity when the solvent approaches the access channel at the active site. Species such as HCO_4^- or HCO_3^- vary in size, charge and redox potential which limits the access of these species to copper in the active site of SOD.

In addition to the above it has been established that SOD-catalyzed peroxidations present dependence on CO_2 rather than HCO_3^-, for example, Mn(II) and Co(II)-catalyzed peroxidations also have this property and involve oxidation of CO_2 to $CO_3^{\cdot -}$. Similarly, the reaction of CO_2 with peroxynitrite produces carbonate radicals, thus demonstrating that there is something in CO_2 that induces it to oxidize to $CO_3^{\cdot -}$.

1.1.1.1. SOD Inactivation by ROS

SOD and catalase enzymes can be affected by ·OH hydroxyl radicals when these are produced by metal-catalyzed decomposition bound to proteins. Therefore, the inactivation of these enzymes by hydroxyl radicals may be due to precursor inactivation by H_2O_2 and other peroxidized molecules. Although catalase is inactivated by ·OH, enzymes such as SOD and glutathione peroxidase are less affected by this radical. Similarly, factors such as ozone, hypochlorite and the ascorbate-Fe(III) system inactivate SOD [7].

A constant source of alkyl-peroxyl radicals employed for studies of enzyme inactivation by free radicals is the pyrolysis of 2,2'-azo-bis-(2-amidinpropane) (ABAP) under aerobic conditions. The radicals produced reduce enzyme activity, which decreases when the enzyme concentration is low. Conversely, when the enzyme concentration is high, the inactivation rate decreases.

Therefore, the order of the process

$$\text{Radical} + \text{Enzyme} \rightarrow \text{inactivation} \tag{13}$$

can be between one and zero. A high reactivity of catalase toward alkyl-peroxyl radicals can be obtained because only a small fraction of this enzyme is needed to trap free radicals. That may be due to its high molecular weight. The high reactivity could be associated with the active catalase molecules and the presence of other proteins, so that the order of the processes could be determined by all the proteins involved in the system. The above indicates that catalase may be inactivated by radicals produced by ABAP cleavage under aerobic conditions, but the radical load is not decisive.

For every six radicals involved in the system, approximately one molecule of SOD loses its activity. Thus, for each reactive radical/enzyme interaction, the activity of the enzyme is reduced by approximately 20%. This was determined from studies performed at high concentrations where the order of the processes tends to zero. As mentioned above, the charges of the radicals produced from ABAP do not affect the high sensitivity of the enzyme, because radicals produced in the pyrolysis of azo-bis-cyanovaleric acid similarly inactivate the enzyme.

The results from the study indicate that under conditions of oxidative stress due to increased free radicals, SOD and catalase enzymes are sensitive to the loss of their activity by oxidized radicals.

Inactivation rates of SOD and catalase are similar when the enzyme concentration is low and when the concentration of the radicals does not depend on their reactivity with the enzyme. Catalase has high reactivity but low sensitivity to alkyl-peroxyl radicals, which means that similar inactivation rates of the two enzymes could be the result of a compensatory effect.

If the loss of enzyme activity is expressed by:

$$- d[\text{Act}]/dt = k_{\text{inact}}[\text{Act}][\text{Radicals}] \qquad \textit{(Equation 5)}$$

the inactivation constant k_{inact} could be related to the process constant (13) by

$$k_{13} = b k_{\text{inact}} \qquad \textit{(Equation 6)}$$

where b is the number of radicals that may react with the enzyme to inactivate it. Since k_{13} and b are higher for catalase than for SOD, then the k_{inact} values for both enzymes are similar.

In studies performed by Escobar *et al.*, 1996 they found, that first-order inactivation rates are achieved at low enzyme concentrations, thus demonstrating that inactivation rate does not depend on enzyme concentration. This is supported by the fact that the inactivation rate of catalase 40nM is not affected by the addition of SOD 4nM. Similarly, the inactivation rate of SOD 4nM is not affected by the addition of catalase 40nM.

Under these conditions, singlet oxygen concentrations in the fundamental state become substrate-independent and are only limited by their unimolecular dissociation. Enzyme inactivation may be represented by the following processes

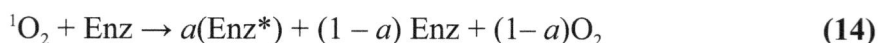

$$^{1}\text{O}_2 + \text{Enz} \rightarrow a(\text{Enz*}) + (1 - a)\,\text{Enz} + (1 - a)\text{O}_2 \qquad \textbf{(14)}$$

Where Enz* is the standard for an inactivated enzyme. Enzyme activity loss (Act) can be expressed as

$$- (d\text{Act})/dt = k_{\text{inact}}[^{1}\text{O}_2]_{\text{ss}}\text{Act} \qquad \textit{(Equation 7)}$$

SOD and catalase can also be inactivated by singlet oxygen. The above was established with $k_{\text{inact}} = a k_{14}$, where k_{inact} values of 2.5×10^7 $M^{-1}s^{-1}$ for catalase and 3.9×10^7 $M^{-1}s^{-1}$ for SOD were obtained. The order of magnitude of these values was the same for amino acids such as tyrosine, histidine, and tryptophan. The singlet oxygen concentration has been estimated to be of the order of 0.2 pM, under conditions of *In vivo* oxidative stress. A half-life of 34 h for SOD and 55h for catalase has been estimated under these conditions.

Enzyme interactions with singlet oxygen may reduce enzyme concentration in the absence of induction. In addition, under oxidative stress conditions the decreases in catalase and SOD levels are similar.

1.1.2. Catalase

Catalase is a protein with group *hemo* that catalyzes the reduction of H_2O_2 to H_2O and O_2 (reaction 15).

$$2\ H_2O_2 \rightarrow 2\ H_2O + O_2 \tag{15}$$

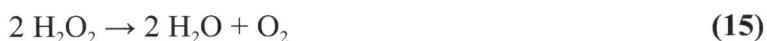

This enzyme has relatively high velocity constants, but its affinity is low; therefore, its role is fundamental in conditions of high hydrogen peroxide concentrations. It has a molecular weight of 240 kD and is made up of four sub-units, each with one *heme* group that is part of its active center [8]. In eukaryotic cells catalase activity is mainly found in peroxisomes, which are the source of most H_2O_2-producing enzymes (Fig. **2**).

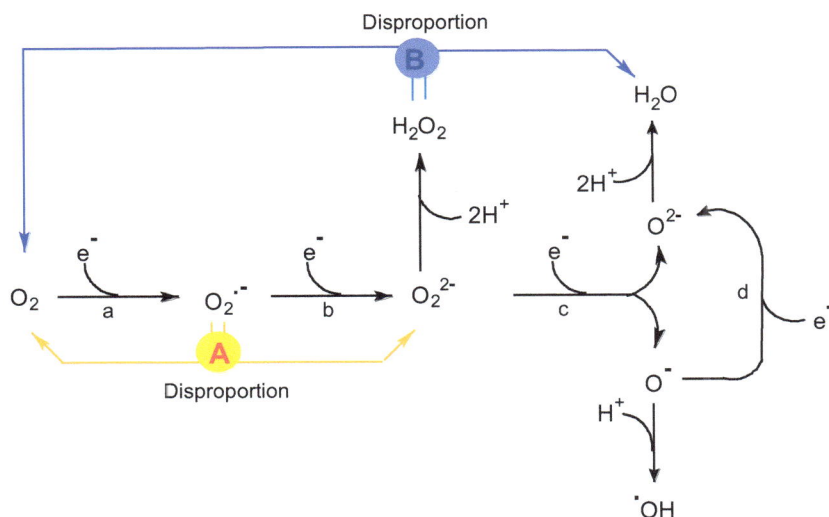

Fig. (2). Superoxide dismutase and Catalase antioxidant enzyme activity. Explanation given in text.

When the O_2 molecule takes an electron (a), the superoxide radical (O_2^-) is formed. A second reduction (b) leads to the anion peroxide (O_2^{2-}) formation which binds easily to protons and is converted to hydrogen peroxide (H_2O_2). The cleavage of the molecule into O_2^- and O^- ions is produced by a third electron (c). The transfer of a fourth electron (d) and the subsequent protonation also converts

O^- to water. The enzyme *Superoxide Dismutase* - SOD breaks down (disproportionates) two superoxide molecules to O_2 and H_2O_2. The H_2O_2 is in turn disproportionate in O_2 and H_2O by the enzyme Catalase *heme*-containing.

1.1.3. Selenium-dependent Glutathione Peroxidase

The selenium-dependent enzyme glutathione peroxidase (SeGSHpx) belongs to peroxidases family which catalyzes with glutathione (GSH) as a cosubstrate the reduction of H_2O_2 and organic hydroperoxides (reactions 16 and 17), which is oxidized into oxidized glutathione (GSSG).

$$H_2O_2 + 2\ GSH \rightarrow GSSH + 2\ H_2O \tag{16}$$

$$ROOH + 2\ GSH \rightarrow GSSG + ROH + H_2O \tag{17}$$

The SeGSHpx is a tetrameric protein of molecular weight 85 kDa that contains four atoms of selenium, bound as selenocysteine, which confer catalytic activity. Unlike catalase, SeGSHpx has a low catalytic activity, but a high affinity for its substrate. This enzyme is found mainly in cytosol and mitochondria of eukaryotic cells [8].

1.1.4. Glutathione Reductase

The SeGSHpx has an absolute GSH requirement to function. The high GSH/GSSG ratio is maintained by the enzyme glutathione reductase. This enzyme catalyzes the oxidized glutathione reduction using redox equivalents in the form of NADPH (reaction 18). Other disulfides can also be reduced by glutathione reductase.

$$GSSG + NADPH + H^+ \rightarrow 2\ GSH + NADP^+ \tag{18}$$

This 120 kDa molecular weight enzyme protein contains two subunits, each with an FAD group in its active center. Its location is cytosolic and mitochondrial.

1.1.5. Glucose-6-phosphate Dehydrogenase

In the pentose phosphate pathway through glucose-6-phosphate, glutathione reductase and SeGSHpx activities are coupled to NADPH production (reaction 19).

$$\text{Glucose-6-phosphate} + NADP^+ \rightarrow \text{6- phosphogluconolactone} + NADPH + H^+ \quad \textbf{(19)}$$

1.1.6. Other Enzymes

There are several peroxidases that have been identified in different biological systems and that have an affinity for H_2O_2. Among these enzymes is peroxyredoxine, which reduces H_2O_2 to H_2O in a redox system in which thioredoxine provides the electrons. These peroxidases have a limited protective effect because many of them transform certain xenobiotics into prooxidants.

In the antioxidant defense of biological systems, enzymes that prevent the formation and/or metabolism of prooxidant species are important. NADPH-quinone oxidoreductase (DT-diaphorase), is an example of these enzymes which catalyzes the divalent reduction of many quinones using NADH or NADPH as an electronic giver and producing stable hydroquinones, which present conjugation reactions to be eliminated. Similarly, hydrolase epoxides, which are found in different cell types, also constitute a primary antioxidant defense system because they react with different epoxy species produced during lipid peroxidation.

1.2. Antioxidant Mitochondrial Defense

Reactive oxygen species - ROS are part of the normal metabolism of cells. When the redox balance is disrupted by excess ROS or by depletion of antioxidants or by both, oxidative stress is generated. Cells must restore homeostatic parameters to reestablish redox balance and counteract the antioxidant effect. Enzymes and low molecular weight antioxidants are the main constituents of the natural antioxidant system. SOD, peroxidase, and catalase are some of the enzymes that trap ROS and low molecular weight antioxidants include glutathione, ascorbate, tocopherols, and phenolic compounds, among others.

A new family of peroxidases that catalyze the reduction of H_2O_2 and alkyl hydroperoxides with the utilization of reducing equivalents donated by thioredoxin is the peroxiredoxins. In mammals, this family has six members found in various subcellular compartments. In mitochondria the enzymes thioredoxin peroxiredoxin III (Fig. **15**, Chapter **3**) together with the enzyme glutathione peroxidase convert H_2O_2 from O^-_2 to water by a reaction catalyzed by MnSOD. H_2O_2 can also diffuse into the cytoplasm. The enzyme phospholipid hydroperoxide glutathione peroxidase is in the cytosol and in the inner membrane of mitochondria and catalyzes phospholipid hydroperoxide production at the expense of GSH, thus being a key enzyme in membrane protection from oxidative stress. The peroxiredoxin III is specific to mitochondria.

In mitochondria by NADP$^+$-dependent isocitrate dehydrogenase and through transhydrogenase-$\Delta\psi$-dependent the NADPH is produced which is required by mitochondrial glutathione. This enzyme is different from that found in the cytosol and because of its isoform, mitochondrial glutathione remains in its reduced state. Catalase, on the other hand, detoxifies mainly H_2O_2 and is found in mitochondrial heart and peroxisomes, providing additional protection [9].

Electron transporters that act as detoxifiers of reactive oxygen species-ROS are found in the phospholipid membrane of the mitochondria. Among them is ubiquinol, called coenzyme Q or QH_2, which has a reducing action for peroxide elimination. Thus, when coenzyme Q is partially reduced or in the form of semiquinone, it is a source of superoxide O^-_2; while when it is totally reduced it is an antioxidant.

Several antioxidant enzymes are in responsible for controlling or eliminating the concentrations of O^-_2 superoxide found in the intermembrane space of the mitochondria, some of them are: Cu,ZnSOD, which is also found in the cytoplasm of eukaryotic cells as an antioxidant for superoxide produced by cytochrome P450 and cytosolic oxidases. The oxygen-generating cytochrome c and the reduced cytochrome c which transfers electrons to the terminal oxidase. Factors such as low pH levels in the intermembrane space induce O^-_2 spontaneous dismutation, due to the release of H^+ during respiration.

In complexes I and III of the mitochondrial electron transport chain, O^-_2 superoxide formation is a non-enzymatic process. When due to the absence of ADP and Pi the electron transport rate is limited and the proton gradient is high, *i.e.* in resting conditions (state 4) O^-_2 formation increases considerably. ROS production can be increased or decreased by electron transport inhibitors depending on their site of action. Under physiological conditions of slow respiration, a high proton motive force is present, therefore electrons accumulate in the Q instead of passing through the electron transport chain to oxygen, which leads to an increase in the ·OH concentration in the resting state and thus increases mitochondrial ROS production.

Between ROS production and membrane potential there is a positive correlation, so that with a slight increase in membrane potential the H_2O_2 production is increased. Likewise, H_2O_2 production can be decreased by 70% with a slight decrease in membrane potential 10mV. During reverse electron transport the high O^-_2 superoxide production by complex I is very sensitive to the membrane potential as well as to a small uncoupling. Therefore, superoxide production by complex I is less sensitive to its membrane potential than to the pH gradient.

In mitochondria, during the oxidation of NADH or succinate without inhibitors no significant amounts of ROS are detected, implying that during electron flow ROS production is not very significant. ROS production decreases when the mitochondrial potential is reduced. This includes chemical uncouplers such as ADP, myxothiazole and malonate (succinate dehydrogenase inhibition); fatty acids can reduce ROS and cause slight uncoupling. The membrane potential decreases when there is a change in respiration from state 4 (resting) to state 3 (active ATP production), so that it has the same effect on the slight uncoupling and ROS production. On the other hand, the proton transport function induced by adenine nucleotide translocase and uncoupling proteins (UCPs) are additional mechanisms by which membrane potential levels can be controlled.

Radical production in complexes I and III is strongly influenced by the proton-motive force. Thus, the activity of uncoupling proteins in the mitochondrial inner membrane influences the free radical production in these complexes. This is due to an increase in proton conductance by superoxide and HNE through the UCP, which leads to a decrease in mitochondrial proton-motive force and thus a strong reduction in ROS production. UCPs decrease the damage caused by mitochondrial ROS production.

In the mitochondrial electron transport chain, the $O^{\cdot-}_2$ superoxide and $HO_2^{\cdot-}$ hydroperoxyl radicals are generated. From the acyl chains of polyunsaturated fatty acids present in the phospholipid membrane these radicals extract hydrogen atoms producing carbon-centered fatty acyl radicals which react with O_2 to form peroxyl radicals. The alkenals produced activate UCPs and ANT expanding the inner mitochondrial membrane proton conductance [9].

The high cellular oxygen content and the high reduced state of the electron transport chain favor superoxide production. In the mitochondrial matrix superoxide reacts with aconitase releasing Fe(II). ·OH hydroxyl radicals can then be formed from H_2O_2 in the presence of Fe(II) by the Fenton reaction. ·OH radicals can also bind to the fatty acyl chains of the phospholipid membrane producing carbon-centered fatty acyl radicals.

When membrane superoxide acts as a hydroperoxyl radical, it can lead to the same effects. Considerable amounts of highly reactive aldehydes are produced by fatty carbon radicals when they lead to self-propagation of the free radical chain reactions. The main products are hydroxyhexenal derived from the *n*-3 fatty acid chain and HNE derived from the *n*-6 fatty acid chain. These compounds attenuate the damage originating from superoxide because they induce the reduction of the proton motive force-Δp and the decrease of superoxide production by the electron transport chain-ETC by activating the uncoupling mediated by UCPs. In contrast, the increase in membrane potential and mitochondrial ROS production is generated when UCPs are inactivated by GDP in mitochondria. The above is consistent with the effect of UCPs on the decrease in ROS production caused by a slight uncoupling. The product of lipid peroxidation 4-hydroxy-trans-2-nonenal - HNE induces mitochondrial uncoupling through adenine nucleotide translocase - ANT and UCP1, UCP2 and UCP3. This is one of the mechanisms by which superoxide activates UCPs [10].

A complex mixture of short-chain aldehydes: 4-hydroxy-2-alkenals such as HNE, 2-alkenals such as 2-hexenal and ketoaldehydes such as malondialdehyde, is generated by the peroxidation of the acyl chains of polyunsaturated fatty acids of phospholipids. Such aldehydes have been shown to have a signaling function, in addition to generating cytotoxic effects associated with oxidative stress.

UCPs may be activated by 2-alkenals, retinoid TTPNB, *trans*-retinoic acid, as well as by double bonds, acyl, or carbonyl groups. Through reaction with proteins and aldehyde group by crossing the double bond C2=C3 and becoming more electropositive by the 4-hydroxy group, the 4-hydroxy-2-alkenals particularly HNE are the most bioactive. For their part, because of the presence of the C=C double bonds and the acyl group, the retinoic acid and TTPNB are activated similarly to HNE.

The 4-hydroxyalkenals react like some free amino acids. However, by addition of Michaelis and Schiff bases most HNE-protein adducts are produced with specific lysine, cysteine, or histidine residues. The presence of groups on the double bonds forms stable covalent adducts generating inter- and intraprotein cross-links. ANT is modified in mitochondria exposed to superoxide and inhibited by HNE.

The physiological function of UCPs: UCP2, UCP3, avian UCP and plant UCP is not well defined. However, taking into account that the magnitude of the proton driving force considerably influences superoxide production in the matrix from complex I and that this superoxide activates UCPs conductance and induces a slight uncoupling, then it can be said that the negative feedback loop in which UCPs intervene consists in that they catalyze the uncoupling originated by the overproduction of superoxide in the matrix. This decreases the proton motive force, thus decreasing the production of superoxide in the ETC and attenuating the damage caused by superoxide, which generates low efficiency of oxidative phosphorylation. The thermogenic function of UCP1 is already well characterized.

1.2.1. Mitochondrial Coenzyme Q. Ubiquinone

The structure of CoQ consists of a quinoid group bound to a hydrophobic end of several 5-carbon isoprene units that varies according to species, *e.g.* 10 isoprene units in humans, 9 in rodents, 8 in *Escherichia coli* and 6 in *Saccharomyces cerevisiae*. Oxidized ubiquinone -UQ and reduced ubiquinol -UQH_2 are the predominant redox forms of CoQ. By addition of two electrons and two protons UQ is reduced to UQH_2 by ubiquinone reductase (reaction 23 and Fig. **3**).

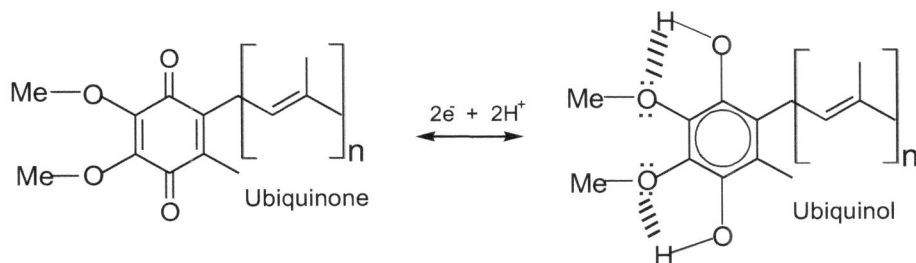

Fig. (3). Coenzyme Q and its redox forms.

The quinol head is more polar and sufficiently hydrophobic to distribute into the phospholipids of the bilayer, although the isoprenoid end has low polarity. Because of the loss of the hydrogen bridge from water on displacement of the quinol to a hydrocarbon environment, the cyclohexane/water partition coefficients of many quinols are lower than those of the corresponding quinones. However, for CoQ because the intramolecular hydrogen bonds between the hydroxy and *methoxy* groups weaken the hydrogen bonds of UQH_2 with water, the *methoxy* groups at positions 2 and 3 decrease this difference.

For both reduced forms of CoQ the octanol/water partition coefficients are similar, as they occur in the quinone/quinol pair due to their hydrogen bridging with octanol. The solubility of the quinone head lipids of UQH_2 and UQ implies that they will partition to approximate degrees within an aprotic medium, the hydrophobic center of the phospholipids of the bilayer, which is analogous to cyclohexane and the membrane closure that is octanol-like at the water-lipid interface. Those CoQ molecules that have a short isoprenoid end are extremely hydrophobic, as indicated by their octanol/water partition coefficient. Therefore, CoQs are not very water soluble and are always found in the phospholipid bilayer.

In the Table **2**, [11], the values for couples O_2/O_2^- and O_2/HO_2^- are based on 1 $M[O_2]$. The values are -330 and -460 mV if 1 atmosphere of O_2 is utilized.

Table 2. Reduction potentials for CoQ and O_2 reactions.

No.	Reaction	$E_{m,7}$ (mV)
20	$UQ + e^- \rightarrow UQ^{\cdot -}$	-240 a -230
21	$UQH^{\cdot} + e^- \rightarrow UQH^-$	190
22	$UQH_2^- + e^- \rightarrow UQH_2$	>850
23	$UQ + 2e^- + 2H^+ \rightarrow UQH_2$	70
24	$O_2 + e^- \rightarrow O^{\cdot -}_2$	-160
25	$O^{\cdot -}_2 + 2H^+ + e^- \rightarrow H_2O_2$	890
26	$HO_2^{\cdot} + H^+ + e^- \rightarrow H_2O_2$	1060
27	$O_2 + H^+ + e^- \rightarrow HO_2^{\cdot}$	-290

Although UQ and UQH_2 are the most stable forms of CoQ, there are other biologically important protonation and redox forms (Fig. **4**, their short-lived pK and $E_{m,7}$ values are given in Table **2**) that are based on their electron transporter and antioxidant activities.

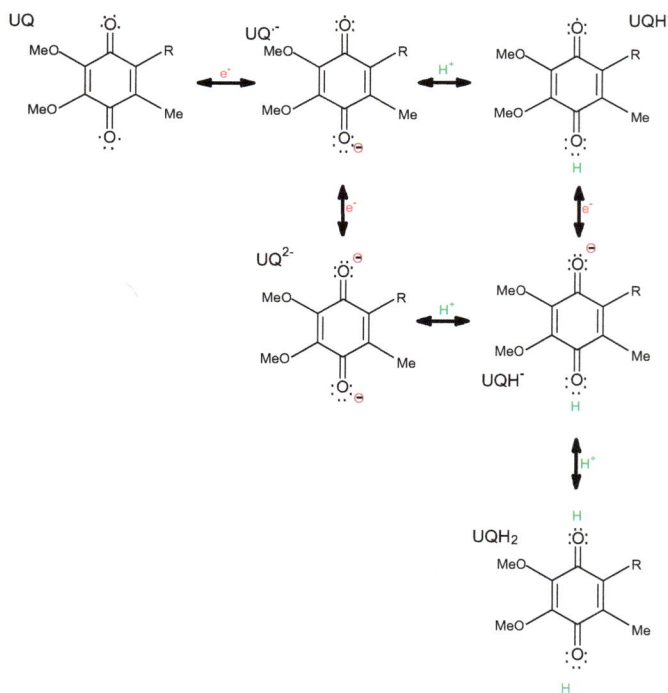

Fig. (4). Principal redox and protonation states of CoQ and its interconversions.

In an aqueous medium, simulating the water/phospholipid interface of the mitochondrial inner membrane, these values were determined. However, *In vivo* the local environment of CoQ may be different. The quinoid group of CoQ may move from the hydrophobic center of the membrane to the vicinity of the water/lipid interface even though it is embedded in the phospholipid bilayer. The pK of the phospholipid bilayer can be increased by changes occurring in the dielectric constant. The reactivity of CoQ on both sides of the membrane can be affected by the pH gradient between the mitochondrial matrix and the cytoplasm being approximately 7.8 - 8 and 7.2, respectively. Furthermore, E_m and pK values could be changed by the binding of CoQ to the enzyme.

In CoQ the electron transfer mechanism occurs in the outer sphere by single electron transfer -SET. UQH_2 is a weak reductant by itself, such that the oxidation of UQH_2 to UQH_2^+ from $E_{m,7}$ has a value of >850 mV. The ubiquinolate monoanion UQH^- is considerably more reduced $E_{m,7} \sim 190$ mV, is produced by UQH_2 on loss of an electron at p$K \sim 11.3$. UQH^- in turn, can donate an electron to the slightly oxidized redox partners. For its part, the ubisemiquinone radical anion $UQ^{\cdot-}$ which is a strong reductant $E_{m,7} \sim -240$ to -230 mV, is generated by the neutral ubisemiquinone radical $UQH^{\cdot}$ on loss of a proton p$K \sim 5.9$. In turn the radical anion ubisemiquinone $UQ^{\cdot-}$ donates an electron to generate UQ (reaction 20 and Fig. **4**).

Only a strongly reduced pair can direct the one-electron reduction of ubiquinone UQ to the ubisemiquinone radical anion $UQ^{\cdot-}$. $UQ^{\cdot-}$ can also be reduced to the UQ^{2-} quinolate dianion which rapidly captures two protons pK of 13.2 and 11.3, respectively. This reduction can be conducted by semi-reduced pairs $E_{m,7} \sim 70$mV. Because of the proximity of the charges stabilizing the ubisemiquinone and quinolate anions, the pK and E_m values of various CoQ intermediates are suitably modified at the enzyme active sites.

In the aprotic lipid bilayer where UQH^- is unstable, this reaction is scarce due to the deprotonation required prior to the transfer of an electron from UQH_2. Similarly, in absence of proton exchange to form the neutral radical $UQH^{\cdot}$, the reduction of UQ to $UQ^{\cdot-}$ is inhibited. However, near the surface of membrane phospholipids, these reactions can occur where proton exchange is possible [11].

Considering that deprotonation is required for UQH_2 to act as a reductant, electron movement does not occur through the UQH_2 and UQ chains but diffusion of UQH_2 does occur. Although the rate of electron transfer from UQH_2 to plastoquinone is scarce in aprotic solvent, UQH_2 can act as a reductant in a protic solvent. Therefore, electron transfer from ubiquinol to ubiquinone does not occur in the phospholipid bilayer but can occur under conditions that allow the formation of UQH^-.

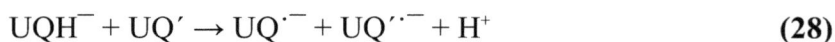

$$UQH^- + UQ' \rightarrow UQ^{\cdot -} + UQ'^{\cdot -} + H^+ \tag{28}$$

In yeast and plant mitochondria in the electron transport reduction is performed by proton pumping NADH dehydrogenase. However, in the mitochondrial transport chain of other cells, CoQ is an important component that receives electrons from several oxidoreductases that reduce UQ to UQH_2. Among these oxidoreductases are: Complex I, Complex II, flavoprotein-ubiquinone electron transfer oxidoreductase, dihydroorotate dehydrogenase and glycerophosphate dehydrogenase.

In mammalian mitochondria, UQH_2 is reoxidized to UQ by complex III and in plants and animals it is carried out by alternative oxidases. The relative rates of electron flow into and out of the CoQ pool influence the fundamental state of UQH_2/UQ. By decreased electron flow out of CoQ through inhibition of complex III or IV, or by elevated electron supply, this redox couple is changed to a lower state. CoQ becomes more oxidized by stimulation of complexes III and IV, by inhibition of UQ reductase and by the reduction in substrate supply. When the mitochondria are changed from the non-phosphorylated to the phosphorylated state, a biologically important redox change of CoQ occurs.

The rate of respiration could be low when the mitochondrion is not producing ATP, for example in state 4, causing the CoQ pool to be lower during unlimited substrate supply. On the other hand, respiration could be fast when the mitochondrion is producing ATP, *e.g.* in state 3, thus causing CoQ pool oxidation. The CoQ pool is reduced by 50-60% in state 3 and by 75-90% in state 4 in the case of isolated mitochondria. However, *in vivo* mitochondria may change these conditions rapidly, *i.e.* they may oxidize or reduce the CoQ pool.

The CoQ concentration in rodent mitochondria is approximately 1 and 5 nmol CoQ/protein, with a higher concentration in skeletal muscle and heart and a lower concentration in brain and liver. The CoQ pool is mobile and easily diffusible in the inner membrane of mitochondria. It is estimated that a CoQ molecule, during respiration in state 3, changes its redox state every 50 ms, due to a rapid interconnection between UQH_2 and UQ that occurs in the mitochondrial inner membrane. In the complex III structure, which has a cavity through which CoQ can enter the active site, this rapid change between bound and free CoQ takes place.

CoQ distribution through the circulation is scarce since the organism must synthesize its own CoQ. From acetyl CoA and through mevalonate, isoprene units are made and then assembled into polyprenylpyrophosphate. The *p*-hydroxybenzoic acid is the precursor of the quinoid head. Through various reactions such as hydroxylation, decarboxylation, and methylation, *p*-hydroxybenzoic acid is conjugated with polyprenylpyrophosphate to produce the CoQ molecule. *In vivo* CoQ half-life is approximately 50-125 h and its breakdown products are excreted in the urine as short-chain phosphorylated derivatives.

1.2.1.1. *Antioxidant Properties of CoQ*

CoQs main antioxidant role is to prevent lipid peroxidation. In lipid peroxidation a ·OH hydroxyl radical initiates a chain reaction by abstracting the H· hydrogen atoms of a phospholipid from the unsaturated fatty acid LH (Chapter 1).

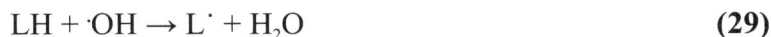

$$LH + \cdot OH \rightarrow L^{\cdot} + H_2O \tag{29}$$

LOO· peroxyl radicals, formed by the reaction of L· carbon-centered radicals and oxygen, may abstract an H· from an unsaturated fatty acid and generate lipid peroxide which propagates the chain reaction:

$$L^{\cdot} + O_2 \rightarrow LOO^{\cdot} \tag{30}$$

$$LOO^{\cdot} + L'H \rightarrow LOOH + L'' \tag{31}$$

LO· alkoxyl radicals, which also extract an H· from fatty acids to form carbon-centered radicals, are formed by the decomposition of lipid peroxide in the presence of Fe^{2+} or Cu^{+}:

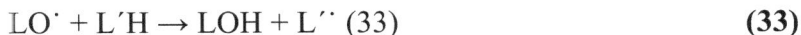

$$LOOH + Fe^{2+} \rightarrow LO^{\cdot} + OH^{-} + Fe^{3+} \tag{32}$$

$$LO^{\cdot} + L'H \rightarrow LOH + L'' \ (33) \tag{33}$$

Antioxidants mainly phenolic HA protectants cleave the chains preventing lipid peroxidation, through the donation of an H· to a carbon-centered radical or oxygen, and thus to produce a non-radical molecule.

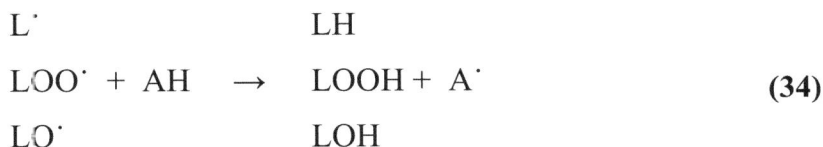

$$\begin{matrix} L^{\cdot} & & LH & \\ LOO^{\cdot} & + \ AH & \rightarrow & LOOH + \ A^{\cdot} & \tag{34} \\ LO^{\cdot} & & LOH & \end{matrix}$$

The propagation of the chain reaction is interrupted by transferring the unpaired electron to antioxidant $A\cdot$. There, the radical character is delocalized on the phenyl ring of radical $A\cdot$ making it a less reactive radical than the original one. Once the $A\cdot$ radical is stabilized when there is a recycling pathway for the radical derived from the antioxidant, it regenerates to its active form or reacts with another radical.

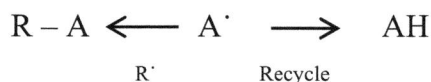

$$R - A \longleftarrow A\cdot \longrightarrow AH$$
$$ R\cdot \qquad Recycle$$

An important chain-breaking antioxidant to prevent lipid peroxidation *In vivo* is Vitamin E, in its predominant α-tocopherol αTOH form. The α-tocopheroyl radical αTO$\cdot$ is formed by the reaction of vitamin E, which is hydrophobic with radicals present in the phospholipid bilayer (reactions 35 and 36). This radical can be recycled back to active αTOH by reductants such as ascorbate (reaction 37). The constants of some of these reactions are presented in Table **3** [11]:

Table 3. Reaction constants for lipid peroxidation.

No.	Reaction	Constant ($M^{-1}s^{-1}$)
29	$LH + \cdot OH \rightarrow L\cdot + H_2O$	5×10^8
30	$L\cdot + O_2 \rightarrow LOO\cdot$	3×10^8
31	$LOO\cdot + LH \rightarrow LOOH + L\cdot$	30–40
32	$LOOH + Fe^{2+} \rightarrow LO\cdot + OH^- + Fe^{3+}$	3.2×10^2
33	$LO\cdot + LH \rightarrow LOH + L\cdot$	4.4×10^6
35	$\alpha TOH + LOO\cdot \rightarrow LOOH + \alpha TO\cdot$	3.3×10^6
36	$\alpha TOH + LO\cdot \rightarrow LOH + \alpha TO\cdot$	1×10^8
37	$UQH_2 + \alpha TO\cdot \rightarrow \alpha TOH + UQH\cdot$	2.2–3.7×10^5
38	$UQH_2 + LOO\cdot \rightarrow LOOH + UQH\cdot$	3.4×10^5

The formation of the ubisemiquinone radical is the central reaction of antioxidant efficacy where this radical is formed by the transfer of an $H\cdot$ atom from UQH_2 to the radical. This reaction is a concerted reaction that occurs in the phospholipid bilayer to avoid sequential transfer. Unlike the sequential transfer of electrons and protons that occur by CoQ. However, sequential electron and proton transfer could occur in an aqueous environment.

$$\begin{matrix} & L^{\cdot} & & & & LH \\ UQH_2 + & LOO^{\cdot} & \rightarrow & UQH^{\cdot} & + & LOOH \\ & RO^{\cdot} & & & & ROH \end{matrix} \qquad (38)$$

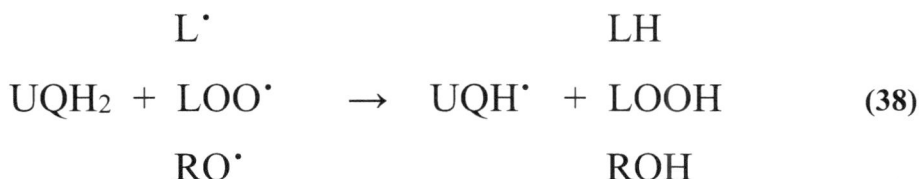

The UQH$^{\cdot}$ radical obtained by donating a hydrogen atom H$^{\cdot}$ may react with another radical to generate UQ, may react with oxygen or be recycled.

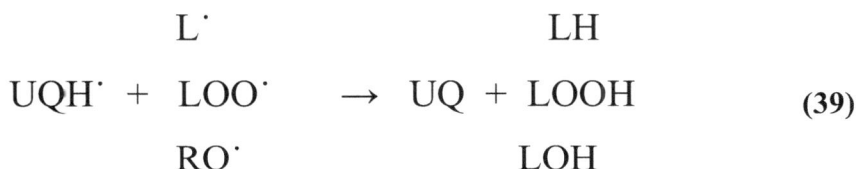

$$\begin{matrix} & L^{\cdot} & & & & LH \\ UQH^{\cdot} + & LOO^{\cdot} & \rightarrow & UQ & + & LOOH \\ & RO^{\cdot} & & & & LOH \end{matrix} \qquad (39)$$

The main function of ubiquinol and ubiquinone is focused on mitochondrial lipid peroxidation when the COQ pool is oxidized. It is important to note that UQ, lacking phenolic hydrogen atoms, does not have antioxidant activity as a chain breaker.

In mitochondria through the reactions of UQH$_2$ and the α-tocopheroxyl radical, vitamin E and CoQ cooperate as antioxidants, to recycle α-tocopheroxyl to its active form (reaction 37). In mitochondria *in vivo*, antioxidant action as chain breakers is difficult as there is 10 times more CoQ than vitamin E. Whereas the main function of UQH$_2$ is as a chain breaker. Ubiquinol, on the other hand, due to the intramolecular hydrogen bridge between the hydroxyl and *methoxy* groups in CoQ, is approximately 10% effective as a chain-breaking antioxidant, in homogeneous solution.

Similarly, αTOH is formed by the reaction of the UQH$^{\cdot}$ radical with αTO$^{\cdot}$.

$$UQH^{\cdot} + \alpha TO^{\cdot} \rightarrow UQ + \alpha TOH \qquad (40)$$

Lipid peroxidation may be prevented by UQH$_2$ reacting directly with a free radical or through the αTOH generation (reactions 37 and 40).

Superoxide anion reacts directly with ubiquinone reducing it to ubisemiquinone in aqueous solution.

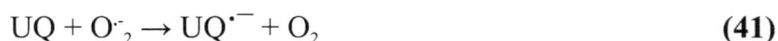

$$UQ + O^{\cdot\,-}_2 \rightarrow UQ^{\cdot\,-} + O_2 \tag{41}$$

Table **4** [11] presents the most important reactions of CoQ with ROS.

Table 4. Reaction constants of CoQ and its derivatives.

Reaction No.	Reaction	Constant $(M^{-1}s^{-1})$
41	$UQ + O^{\cdot\,-}_2 \rightarrow UQ^{\cdot\,-} + O_2$	5×10^6
43	$UQ^{\cdot\,-} + UQ^{\cdot\,-} \rightarrow UQ + UQ^{2-}$	8.4×10^4
44	$UQH^{\cdot} + UQH^{\cdot} \rightarrow UQ + UQH_2$	4.8×10^7
45	$O^{\cdot\,-}_2 + O^{\cdot\,-}_2 + 2H^+ \rightarrow H_2O_2$	1.8×10^5
46	$UQ^{\cdot\,-} + O_2 \rightarrow UQ + O^{\cdot\,-}_2$	2×10^8
47	$UQH_2 + O_2 \rightarrow UQ + H_2O_2$	1.5

UQH^- is also oxidized to $UQ^{\cdot}$ by the superoxide anion

$$UQH^- + O^{\cdot\,-}_2 + H^+ \rightarrow H_2O_2 + UQ^{\cdot\,-} \tag{42}$$

Thermodynamically, the direct reaction of the superoxide anion with UQH_2 is not favorable, and its reaction constant is low (reaction 54).

Superoxide also exists as its acid conjugate, the hydroperoxyl radical $HOO^{\cdot}$, pK 4.8 in the phospholipid bilayer where it can react with UQ and UQH_2:

$$Q + HO^{\cdot}_2 \rightarrow O_2 + UQH^{\cdot} \tag{48}$$

$$UQH_2 + HO^{\cdot}_2 \rightarrow H_2O_2 + UQH^{\cdot} \tag{49}$$

In the phospholipid bilayer $HOO^{\cdot}$ and UQH_2 are found. In hydrophobic medium the pK of $UQH^{\cdot}$ is high and $HOO^{\cdot}$ exhibit higher oxidation than $O^{\cdot\,-}_2$ (Table **2**). These reactions are also important *in vivo*.

$UQ^{\cdot\,-}$ ubisemiquinone radicals are formed from UQH_2 and UQ in the phospholipid bilayer where they remain as neutral $UQH^{\cdot}$ radicals, whereas in aqueous solution $UQH^{\cdot}$ could be reduced to $UQ^{\cdot\,-}$ at pK 5.9. The $UQ^{\cdot\,-}$ radical by donation of an H· reacts and traps another radical. Thus, the semiquinone radical $UQ^{\cdot\,-}$ reacts with oxygen to form superoxide, in a prooxidant action (reaction 57).

$$R^{\cdot} + UQH^{\cdot} \rightarrow RH + UQ \tag{50}$$

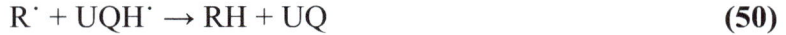

In another way, two ubisemiquinone radicals can also be disproportionated to produce UQ and UQH_2. There are three possibilities: reactions 44, 51 and 52.

$$UQH^{\cdot} + UQ^{\cdot-} + H^{+} \rightarrow UQH_2 + UQ \tag{51}$$

$$UQ^{\cdot-} + UQ^{\cdot-} + 2H^{+} \rightarrow UQH_2 + UQ \tag{52}$$

The reaction between the two neutral radicals of semiquinone or the reaction between the anion and the protonated form are faster, whereas due to repulsion the reaction between the two anions is slower (Table **4**).

In the phospholipid bilayer, the reaction between the two neutral ubisemiquinone radicals is predominant. The ubisemiquinone radical $UQ^{\cdot-}$ formed by the inactivation of reactive oxygen species -ROS, is converted to UQ and UQH_2 and through the respiratory chain UQ can be recycled to UQH_2 (Fig. **5**). This figure presents the CoQ interaction reactions at the phospholipid bilayer of the mitochondrial inner membrane, where UQH_2 deprotonation does not occur, and the water/lipid interface, where protonation and deprotonation of the CoQ forms take place. In the same figure the antioxidant and prooxidant reactions of CoQ are presented.

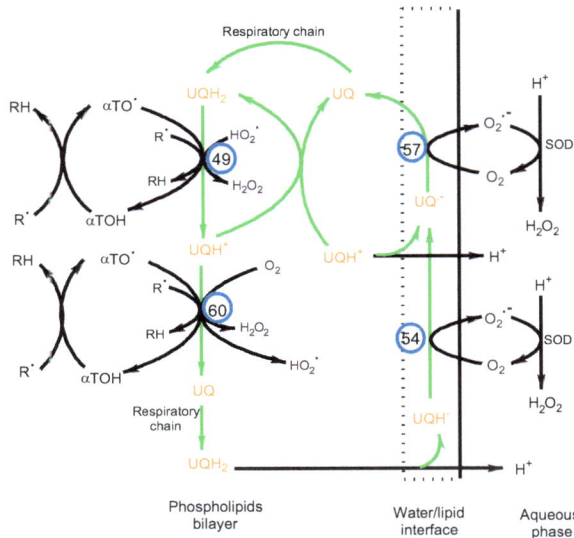

Fig. (5). CoQ mitochondrial reactions as an antioxidant and prooxidant. Reactions are explained in the text.

Unlike vitamin E tocopheroxyl radicals, ubisemiquinone radicals are not reduced by soluble electron donors such as ascorbate and glutathione [11].

The understanding of how endogenous UQ in mitochondria acts as an antioxidant is important. In human pathologies, exogenous UQs such as idebenone, coenzyme Q_{10} and mitoquinone -MitoQ are also employed in therapy to reduce oxidative damage.

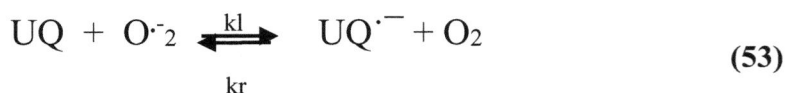

$$UQ + O^{\cdot-}_2 \; \underset{kr}{\overset{kl}{\rightleftharpoons}} \; UQ^{\cdot-} + O_2$$

(53)

The knowledge of the UQ/UQH_2 and $O^{\cdot-}_2/HOO^{\cdot}$ reactions allows the understanding of the antioxidant properties of endogenous and exogenous UQ. It could be determined that the local environment influences the reaction constants. In the case of UQ this may vary from the hydrophobic center of the membrane of endogenous UQ to aqueous surroundings for therapeutically utilized UQ. Thus, for UQ interaction with $O^{\cdot-}_2$ the reaction constants alter with the local environment. The $O^{\cdot-}_2$ dismutation rates and $UQ^{\cdot-}$ influence the position of this equilibrium. For its part, the acid conjugate of $O^{\cdot-}_2$ the hydroperoxyl radical $HOO^{\cdot}$; $pk_a = 4.7$, may diffuse into this region and across the membrane, when the superoxide is removed from the hydrophobic center of the membrane.

In vivo, endogenous, and exogenous UQ and UQH_2 may be present in hydrophobic aqueous, or intermediate environments. The reactivity of UQ/UQH_2 and $O^{\cdot-}_2/HOO^{\cdot}$ in these conditions was evaluated by Maroz *et al*, 2009 [12], who, using pulse radiolysis, calculated the reaction constants in three solvent systems: water, *n*-hexane, and methanol. The UQ reaction with $O^{\cdot-}_2$ was fast (Table **4**) and this may allow UQ to catch $O^{\cdot-}_2$ *in vivo*. The reaction rate is near the limit of diffusion in *n*-hexane, this suggests that in the hydrophobic center of biological membranes where there is a higher concentration of endogenous UQ, the reaction is particularly fast. In the protein complexes of the respiratory chain, $O^{\cdot-}_2$ is formed by the reaction of oxygen with the electron transporter, whereas the diffusion of $O^{\cdot-}_2$ to the center of the membrane from the aqueous phase is low.

A Lot of the $O^{\cdot-}_2$ can be formed near electron donors and thus rapidly converted to $HOO^{\cdot}$. Exogenous UQ such as MitoQ is found near the membrane surface, a small amount is found in the aqueous phase and most endogenous UQ is found in the hydrophobic center. UQ can react with $O^{\cdot-}_2$ in the aqueous phase and at the membrane/water interface, such that through superoxide dismutase, UQ can complete the degradation of $O^{\cdot-}_2$. So, it could be said that the factors that determine the balance between ubisemiquinone and O_2 with UQ and $O^{\cdot-}_2$ are mainly the rate of ubisemiquinone radical dismutation and O^{-}_2 dismutation by SOD.

The acid conjugate of O^{-}_2 the peroxyl radical $HOO^{\cdot}$ is very reactive $E^{\circ}(HOO^{\cdot}, H^+/H_2O_2) = 1.44V$) and very easily permeates membranes, which is the main cause of O^{-}_2 oxidative damage. The extraction rate of unsaturated fatty acids by $HOO^{\cdot}$ is slow $1\text{-}3x10^3 M^{-1}s^{-1}$, constitutes a fundamental step in the initiation of lipid peroxidation (reaction 59) and is like the reaction rate of $HOO^{\cdot}$ with UQH_2. UQH_2 is not effective in preventing lipid peroxidation by $HOO^{\cdot}$ when an excess of UQH_2 is present in biological membranes. The reaction constant of endogenous α-tocopherol with $HOO^{\cdot}$ is $2 \times 10^5 M^{-1}s^{-1}$ or faster than the reaction constant of $HOO^{\cdot}$ with UQH_2. This means that the reaction with α-tocopherol is more efficient than UQH_2 in preventing the initiation of lipid peroxidation by $HOO^{\cdot}$ in membranes. However, in the mitochondrial membrane when α-tocopherol reacts with a radical generated by UQH_2 the α-tocopheroxyl radical is formed, thus contributing to $HOO^{\cdot}$ detoxification. For its part, the reaction of idebenol and mitoquinol with carbon center radicals and oxygen is very fast $1\text{-}2x10^8 \ M^{-1}s^{-1}$, unlike the low reactivity with $HOO^{\cdot}$

The $UQH^{\cdot}$ radical can be formed by the protonation of $UQ^{\cdot -}$ coming from the reaction of UQ with O^{-}_2 and by the reaction of UQH_2 with carbon-oxygen center radicals, including α-tocopheroxyl. However, it is necessary that $UQH^{\cdot}$ present rapid dismutation to UQ and UQH_2, $4.8x10^7 M^{-1}s^{-1}$ (reaction 44) for these antioxidants to be effective. Most of the UQ thus formed could be reduced back to UQH_2 by the mitochondrial respiratory chain and thus maintain the antioxidant cycle (Fig. **5**).

In the presence of oxygen, UQH_2 spontaneously autooxides itself by the global reaction (reaction 47). Autoxidation is initiated by the ubiquinolate anion reactions (reactions 46 and 54), considering that the reaction between UQH_2 and oxygen is scarce (Table **4**).

$$UQH^{-} + O_2 \rightarrow UQ^{\cdot -} + O^{-}_2 + H^+ \qquad \textbf{(54)}$$

As represented in reaction 7 of Chapter 1 and in reaction 35 of Chapter 3, the O^{-}_2 superoxide formed is dismutated to H_2O_2. Considering the above, ubiquinolate anion autoxidation does not occur at pH below 6, is very slow at pH below 8 and is only significant at pH close to 9. However, under high pH conditions, the disproportionation of ubiquinone to UQ and UQH_2 is slow leading to increased O^{-}_2 formation due to the long lifetime of ubisemiquinone. As autoxidation depends on disproportionation then UQH_2 oxidation in the phospholipid bilayer is not likely *via* this pathway. However, autoxidation can occur near the membrane surface where high-water infiltration to the center of the membrane may occur.

Through electron transfer from the ubiquinolate anion to ubiquinone, the ubisemiquinone radical is formed (reaction 28). Similarly, ubiquinol oxidation is mediated by the presence of metal ions such as ferric or cupric for the oxidation of ubiquinolate to ubisemiquinone. These reactions may occur under the same conditions as the spontaneous oxidation of oxygen, considering that the disproportionation of the quinolate anion is necessary.

The overall reaction described in reaction (47) is generated when the $O^{\cdot-}_2$ superoxide produced by UQH_2 autoxidation is dismutated to H_2O_2 (reaction 51). Conversely, a chain autoxidation can occur, when only one quinol oxidation is initiated, because through the formation of ubisemiquinone radicals generated by $O^{\cdot-}_2$ superoxide reaction with UQ or UQH_2, the superoxide acts as a transporter to propagate the UQH_2 autoxidation:

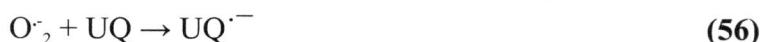

$$UQH_2 + O^{\cdot-}_2 \rightarrow UQ^{\cdot-} + H_2O_2 \qquad (55)$$

$$O^{\cdot-}_2 + UQ \rightarrow UQ^{\cdot-} \qquad (56)$$

Through these reactions UQH_2 rapidly autoxidizes in water, however, this autoxidation can be stopped when the superoxide propagation chain is inhibited by adding SOD. However, some of the superoxide involved in the propagation of autooxidation is inaccessible to SOD since autooxidation of UQH_2 in yeast membrane and mitochondria is independent of SOD.

The redox potentials for the $UQ/UQ^{\cdot}$ and $O_2/O^{\cdot-}_2$ couples are similar (Table **2**). This means that the ubisemiquinone radical formed by transferring an electron from $O^{\cdot-}_2$ superoxide to UQ may in turn react with O_2 to produce $O^{\cdot-}_2$ superoxide.

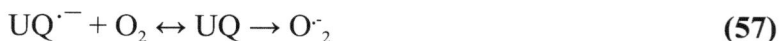

$$UQ^{\cdot-} + O_2 \leftrightarrow UQ \rightarrow O^{\cdot-}_2 \qquad (57)$$

This reaction is diverted to the right by the consumption of $O^{\cdot-}_2$ superoxide thus reducing the ubisemiquinone radical concentration through the formation of H_2O_2 catalyzed by SOD. The above generates that in an oxygen environment, ubisemiquinone reactions are sensitive to SOD. For example, considering that there is a direct reaction between $UQ^{\cdot-}$ ubisemiquinone and cytochrome *c*, then this reaction is SOD sensitive [11].

UQH^- formation is limited in the phospholipid bilayer. Therefore, if the ubisemiquinone radical that is generated from reaction with a free radical, then UQH_2 autooxidation occurs. $HOO^{\cdot}$ acts as a transporter in the autooxidation of chain propagation, in the phospholipid bilayer:

$$R^{\cdot} + UQH_2 \rightarrow UQH^{\cdot} \tag{58}$$

$$UQH^{\cdot} + O_2 \rightarrow UQ + HOO^{\cdot} \tag{59}$$

$$HOO^{\cdot} + UQH_2 \rightarrow UQH^{\cdot} + H_2O_2 \tag{60}$$

When $UQ^{\cdot-}$ ubisemiquinone radical generation, caused by oxidative damage, is faster than the disproportionation of this to UQ and UQH_2, or when they are trapped by chain termination reactions, then UQH_2 is consumed by self-propagation of the chain reaction of autooxidation. When $HOO^{\cdot}$ diffuses into the matrix outside the membrane, it loses a proton and *via* MnSOD is converted to H_2O_2 or reacts with cytochrome *c* or Cu,ZnSOD when it enters the intermembrane space.

1.3. Non-enzymatic Traps

Besides the enzymes described above, there is another line of antioxidant defense that works without enzymatic intervention, trapping free radicals that escape from antioxidant enzymes. Within this group are various proteins and molecules of low molecular weight such as glutathione, vitamin C and phenolic compounds such as flavonoids.

1.3.1. Proteins

Through *Fenton* or *Haber-Weiss* reactions, $^{\cdot}OH$ hydroxyl radicals are produced with the participation of transition metals such as iron and copper (reactions 33 and 39 chapter 3). They also participate in free-radical reactions in which they turn little reactive species into more reactive ones. However, when these metals are bound to proteins, they can hardly carry out this catalysis. There are several proteins that are capable of binding to metals and therefore reduce the levels of free metal ions; these proteins are also considered antioxidant defense mechanisms.

Ferritin and transferrin are proteins that maintain low intracellular and extracellular iron concentrations. Ferritin is involved in the intracellular storage of iron and has 24 subunits each with 18.5 kDa molecular weight. This protein be able to store up to 4,500 iron atoms, which are in the internal cavity and result from the different subunits association. The antioxidant capacity of ferritin depends on its level saturation with iron that is present, so that when it is partially saturated, it acts as a powerful antioxidant in the plasma when it traps iron from it, and when it is totally saturated it can release it by becoming a prooxidant.

Transferrin is an 80kDa glycoprotein that transports iron in plasma. Each transferrin molecule can bind up to 2 atoms/gram of iron. Like ferritin, this protein can function as a prooxidant when fully loaded with iron. Ceruloplasmin and albumin transport copper in the plasma and prevent the hydroperoxides decomposition to free radicals. The first is a 130 kDa protein that can transport either 6 or 7 copper ions per molecule. Ceruplasmin is also capable of oxidizing Fe^{2+} to Fe^{3+}, thus preventing Fe^{2+} to catalyze free-radical reactions. Albumin is a protein with a 69 kDa molecular weight, and among its functions is the osmotic pressure regulation and different types of molecules transport in the plasma. This protein, in concentrations lower than normal physiological concentrations, can inhibit the lipid peroxidation stimulated by copper. Albumin inhibits the hydroxyl radical generation in systems containing copper and H_2O_2 ions and can trap hydroxyl and peroxyl radicals. It can also bind to free fatty acids by protecting them from peroxidation; however, in this case, its effect on iron-stimulated peroxidation is minimal [8].

1.3.2. Glutathione

The tripeptide L-γ-glutamyl-L-cysteinyl-glycine, or GSH, is the most important thiol compound in plants and animals. It has a molecular weight of 307, reaches concentrations up to 10mM and its peptide-γ-bond protects it from degradation by aminopeptidases. In cells, tissues and plasma, glutathione is found in various forms. Glutathione is known as a cellular thiol buffer-redox, capable of maintaining the thiol/disulfide redox potential, because it has no toxicity associated with cysteine. Oxidized glutathione-GSSG also known as glutathione disulfide is an oxidation product. Glutathione-cysteinyl -GSSR, is biologically important in proteins. Therefore, proteins are glutathionylated or thiolated.

Glutathione GSH is synthesized from L-cysteine, glycine and L-glutamate in two-step reactions catalyzed by γ-glutamyl cysteine synthase and glutathione synthase. Through the enzymatic centers cysteinyl-glycine dipeptidase and γ-glutamyl-transpeptidase the transformation of the constituent amino acids takes place. Disulfide and thioether and thioester formation are redox reactions related to these enzyme centers [13] (Fig. **6**).

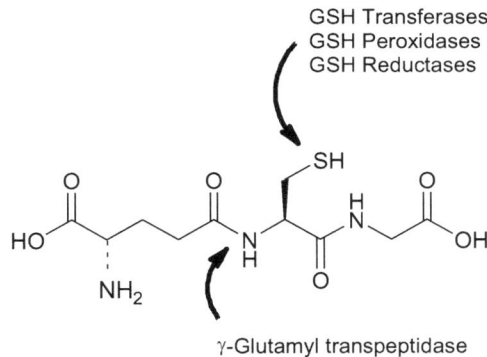

Fig. (6). Glutathione (GSH).

Several types of GSH peroxidases and GSSG reductases catalyze the redox reactions, whereas glutathione transferase-GST catalyzes the formation of thioesters (Fig. 7). In this figure glutathione (Glu-Cys-Gly sequence) contains a γ-peptide bond between Glu and Cys. Cysteine residue thiol group is redox-active. During oxidation two GSH molecules bind to the disulfide bridge.

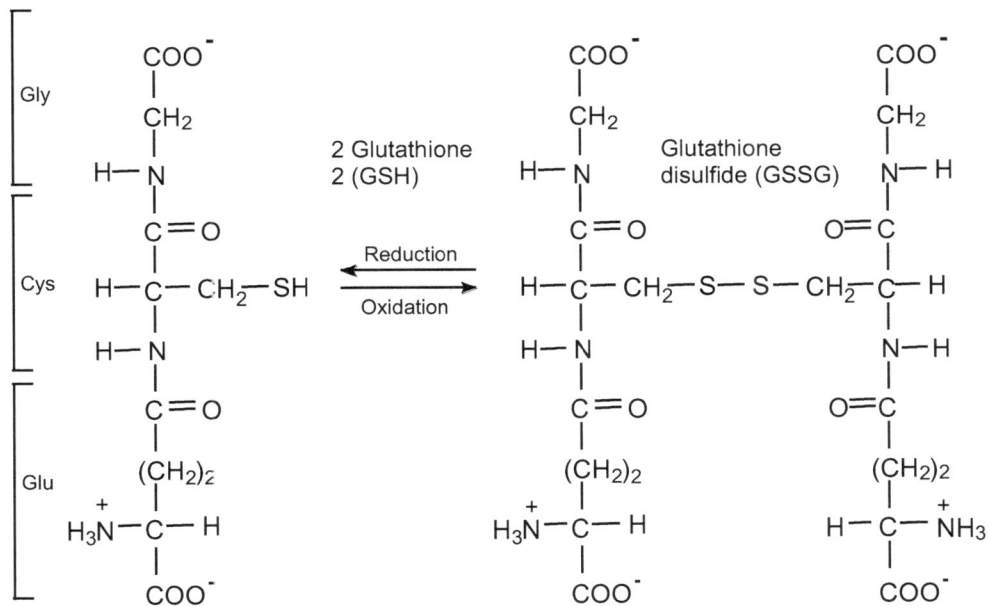

Fig. (7). Redox reactions. Explication given in text.

Glutathione can react with oxygen free radicals in different ways. Firstly, by glutathione peroxidase action it can reduce species such as H_2O_2 or other organic peroxides, oxidizing to GSSH (reactions 18 and 19). Secondly, through hydrogen atom transfer and thiyl radical formation that transforms into GSSG, it reacts with free radicals such as $\cdot$OH, O^-_2 and RO$\cdot$ directly. Thirdly, through reactions catalyzed by glutathione transferases, it reacts with electrophiles to form covalent adducts.

Glutathione levels also vary according to physiological conditions. It has been shown that GSH concentration decreases with age, possibly due to a decrease in GSH replacement, due to increased utilization, degradation, reduced biosynthesis, and increased oxidation rate. GSH loss and other cellular thiols favors lipid peroxidation and tissue damage.

Inhibition of lipid peroxidation by GSH appears to be related to vitamin E regeneration in which it is involved (reaction 61). In the reaction, a thiyl radical is produced, which can be combined with another thiyl radical (GS$\cdot$) to form GSSG, which is reduced to GSH by glutathione reductase (reactions 62 and 18 respectively).

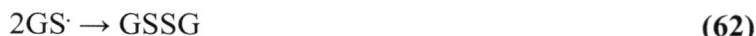

$$\text{Vitamina E}\cdot + \text{GSH} \rightarrow \text{Vitamina E} + \text{GS}\cdot \qquad\qquad \textbf{(61)}$$

$$2\text{GS}\cdot \rightarrow \text{GSSG} \qquad\qquad \textbf{(62)}$$

1.3.3. Vitamin C

Vitamin C, or ascorbic acid, is a molecule that has been found intracellularly and extracellularly in most biological systems. In plasma, it is the water-soluble antioxidant that has the greatest protective effect against lipid peroxidation. At physiological pH, the predominant form is ascorbate anion (AH-) considering that the pK of ascorbic acid is 4,25.

As an antioxidant, ascorbate anion reacts directly with superoxide, hydroxyl, and lipid hydroperoxide radicals. When ascorbate reduces these free radicals, it becomes dehydroascorbate (A) through the free radical intermediate formation, semideshydroascorbate (A$\cdot$) (**Fig. 8**). Dehydroascorbate is an unstable molecule and can break down into a complex pathway leading to both oxalic and L-threonic acid production.

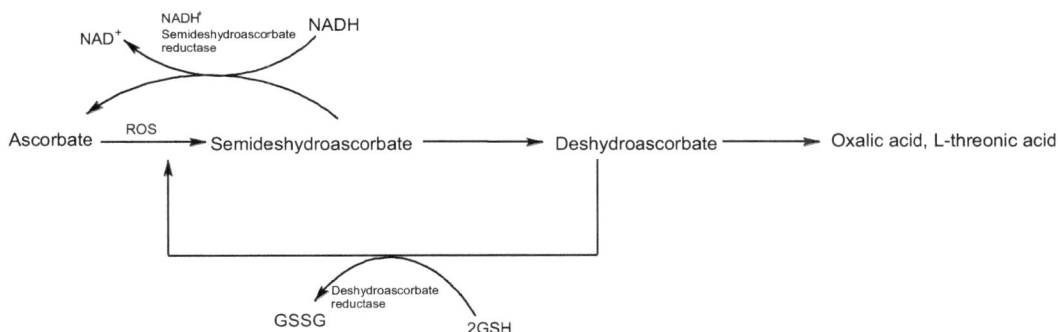

Fig. (8). Ascorbate oxidation by reactive oxygen species (ROS), regeneration and decomposition.

However, ascorbic acid is regenerated from dehydroascorbate by dehydroascorbate reductase utilizing glutathione reduced oxidizing it to GSSH, or from semi-hydroascorbate by NADH-semideshydroascorbate reductase oxidizing NADH to NAD$^+$. It is believed that semideshydroascorbate can also be reduced by glutathione to ascorbate resulting in a thiyl radical (reaction 63).

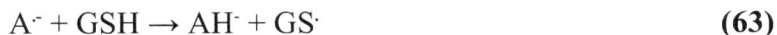

$$A^- + GSH \rightarrow AH^- + GS^- \tag{63}$$

Other important function of vitamin C is to restore the antioxidant properties of vitamin E. In this case the ascorbate is oxidized by reducing the tocopheryl radicals (vitamin E$^-$) originating in the reactions of vitamin E with free radicals (reaction 64).

$$Vitamin\ E^- + AH^- \rightarrow Vitamin\ E + A^- \tag{64}$$

Ascorbate, under certain conditions can also function as a prooxidant. At high concentrations (≈ 1 mM) and with transition metals, this antioxidant can induce the oxygen free radicals generation by its ability to reduce metal ions, which are involved in hydroxyl radical formation reactions [8].

2. SECONDARY ANTIOXIDANT DEFENSE SYSTEMS

2.1. ENZYMES

Antioxidant defense systems are not always 100% effective and intracellular components present oxidative damage; however, cells have another series of enzymes that are able to repair and/or eliminate the effects resulting from protein and lipid damage.

2.1.1. Protein-specific Oxidoreductases

There are numerous enzymes that catalyse redox reactions of the sulfhydryl groups of proteins, including those that reduce disulfide bridges. The latter contribute to the antioxidant defense of the cells, since they reduce the mixed disulfide bridges formed through the oxygen-free radicals' action. These enzymes include the couple thioredoxine reductase and glutarredoxine.

2.1.2. Proteases

Proteolytic enzymes act as secondary defense systems because they degrade many oxidatively modified and damaged proteins and thus prevent cell accumulation. Both prokaryotes and eukaryotes show an increase in proteolytic susceptibility when they present oxidative stress.

Macroxyproteinase (MOP) is a high molecular weight protein complex involved in the non-lysosomal and ATP/ubiquitin independent of oxidation-modified proteins degradation. Denaturation by proteins oxidative modifications could represent a signal for proteins intracellular proteolysis. The hydrophobic remains of undamaged proteins are located inside their three-dimensional structure. However, during oxidative modifications the partial denaturation of these proteins causes the exposure of these residues to the outside of the proteins, increasing their hydrophobicity. Such conformation provides peptide bonds very susceptible to MOP hydrolysis.

2.1.3. Glutathione Peroxidase not Dependent on Selenium

In the metabolism of lipidic hydroperoxides, non-selenium dependent glutathione peroxidase (GSHpx) is the main enzyme involved in hydroperoxide reduction to the respective alcohols. This cytosolic enzyme, unlike SeGSHpx, is not selenium dependent and does not metabolize H_2O_2 but present specificity for low molecular weight organic hydroxyperoxides. GSHpx has low activity against the hydroperoxides embedded in the membranes; therefore, its antioxidant effect depends on the release of these hydroperoxides from the membranes where they are found. It has been proposed that phospholipase A_2 facilitates the activity of GSHpx because it releases peroxidated fatty acids from the membrane phospholipids.

In addition to this GSHpx, another peroxidase is found in mammals that catalyzes the lipid hydroperoxides direct reduction without the intervention of phospholipase A_2. This is a small 23 kDa protein containing selenium and is called phospholipid hydroperoxide glutathione peroxidase.

2.1.4. Phospholipases

The removal of fatty acids from the membrane are mainly removed by the enzyme phospholipase A_2. There are various isoenzymatic forms of phospholipase A_2 and they have a role important in the metabolism and exchange of membrane phospholipids. The activity of phospholipase A_2 is essential for the acylation-desacilation reactions involved in the *of novo* synthesis of specific phospholipidic types. This phospholipase has high specificity on oxidized phospholipids and functions in response to membrane alterations. Membrane fluidity, as well as the balance in membrane fatty acid composition, can be realized by catalyzing transacylation reactions.

Phospholipase C also has a high activity on oxidized substrates. The sequential action of this phospholipase and diacylglycerol lipase may also be involved in membrane restructuring after oxidative damage.

2.2. Non-enzymatic Traps

There are several small molecules that non-enzymatically react with free radical intermediates. These molecules include vitamin E, various carotenoids, ubiquinol or bilirubin.

2.2.1. Vitamin E

Vitamin E includes a series of compounds called tocopherols, which are closely related to each other. Among these, α-tocopherol has the highest antioxidant activity. In many cells, vitamin E is found in the membrane. Due to its lipophilic character, the tocopherol molecule reacts with reactive oxygen species -ROS such as peroxyl radicals, converting them into lipid hydroperoxides by donating a hydrogen atom (reaction 65). The hydroxyperoxides that are subsequently formed are eliminated by the joint action of the GSH peroxidases and phospholipases described above. In this way, the chain reactions which propagate lipid peroxidation are interrupted by vitamin E. [14].

$$\text{Vitamin E} + \text{ROO}^{\cdot} \rightarrow \text{Vitamin E}^{\cdot} + \text{ROOH} \qquad \textbf{(65)}$$

The tocopheroxyl radical (vitamin E$^{\cdot}$) that originates produces a stable adduct by its reaction with another radical (reaction 66) or is reduced by the ascorbate-GSH redox couple found in cytosol (reactions 61 and 64).

$$\text{Vitamin E}^{\cdot} + \text{ROO}^{\cdot} \rightarrow \text{ROO-vitamin E} \tag{66}$$

α-tocopherol acts as a powerful lipophilic antioxidant and suppressor of oxidative damage in biological membranes, lipoproteins and tissues, by removal of ROS such as hydroxyl radical, superoxide and singlet oxygen.

The active site of vitamin E is found on the chromanol ring is at the 6-hydroxyl position, which is in the membrane near the polar surface, while the fityl chain does so together with the phospholipids in their non-polar region. The α-tocopherol, eliminates the peroxyl radical by hydrogen atom transfer:

$$\text{α-TOH} + \text{ROO}^{\cdot} \rightarrow \text{α-TO}^{\cdot} + \text{ROOH} \tag{67}$$

This reaction occurs quickly. Now tocopherol has lost a hydrogen atom, turning into tocopheryl radical (α-TO$^{\cdot}$). Although this radical is not as reactive, it is recycled through the action of two possible molecules, ascorbic acid, and coenzyme Q, through reactions 37, 40, 68 and 69.

$$\text{α-TO}^{\cdot} + \text{ascorbate} \rightarrow \text{Semideshydroascorbate} + \text{tocopherol-OH} \tag{68}$$

Considering the hydrosoluble ascorbate and the fat-soluble vitamin E; this is found closely aligned with phospholipid fatty tails in the membrane and the chromanol head near the membrane surface. Thus, when a radical is formed α-TO$^{\cdot}$, it is projected outside the apolar region and located in the polar zone, interacting with vitamin C which is in the aqueous region, regenerating the vitamin E chromanol region. Another pathway for the regeneration of the radical tocopheryl is:

$$\text{α-TO}^{\cdot} + \text{CoQH}_2 \rightarrow \text{Tocopherol-OH} + \text{CoQH}^{\cdot} \tag{69}$$

the radical is regenerated in the electron transport chain of the mitochondria.

F_2-isoprostanes, is a biomarker of free radical-mediated lipid peroxidation. Roberts *et al.* in 2007 [15], evaluated the effect of vitamin E (*RRR-α*-tocopherol) dosage on the decrease in plasma F_2-isosprostanes concentrations. They found that plasma concentrations of F_2-isosprostanes at week 0 were 61.7 ± 6.5 pg/ml, at 16 weeks of vitamin E dose administration, the concentration of F_2-isosprostanes was 36.3 ± 2.6 pg/ml and up to weeks 18 and 20 the levels remain suppressed ($p < 0.005$). The antioxidant activity of vitamin E on lipid peroxidation was demonstrated in these studies.

In turn, at week 20, plasma concentrations of vitamin E increased significantly in contrast to week 0. However, there was no significant difference between weeks 0 and 20 in plasma concentrations of total serum cholesterol, HDL cholesterol, LDL cholesterol and triglycerides (Table **5**).

Table 5. Plasma lipid and vitamin E levels reported from [15].

	Study start (mg/dl)	Study end (mg/dl)	*p*
Cholesterol	256±23	242±22	N.S.[a]
Cholesterol LDL	158±23	141±26	N.S.
Cholesterol HDL	58±8	58±6	N.S.
Triglycerides	193±78	211±81	N.S.
Vitamin E	1.48±0.24	4.01±0.81	<0.05

[a]N.S. not significant

2.2.2. Carotenoids

Most of these conjugated polyenes have antioxidant activity. β-carotene, which is a vitamin A precursor, is found in the membranes of various tissues in high concentrations. This carotenoid, in addition to trapping singlet oxygen, reacts with the peroxyl radicals generated during lipid peroxidation to form radicals centered in the carbon of stable resonance which, in turn, react with other peroxyl radicals to form a non-radical free compound (reactions 70 and 71). Thus, like vitamin E, β-carotene works as an inhibitor of membrane lipoperoxidation propagation. These reactions occur more easily with low oxygen concentrations and complement the action of vitamin E, which reacts more efficiently with high oxygen concentrations.

$$\beta\text{-Carotene} + ROO^{\cdot} \rightarrow ROO\text{-}\beta\text{-carotene}^{\cdot} \qquad \textbf{(70)}$$

$$ROO\text{-}\beta\text{-carotene}^{\cdot} + ROO^{\cdot} \rightarrow ROO\text{-}\beta\text{-carotene-OOR} \qquad \textbf{(71)}$$

A molecule of β-carotene reacts with several peroxyl radicals to form carbon-centered radicals, which are then blocked by interacting with other peroxyl radicals. β-carotene, like vitamin C, also works as a prooxidant. With partial oxygen pressures below 150 torr it is an excellent free radical trapper, while with very high oxygen pressures it shows autocatalytic prooxidant effects.

Many of the antioxidants discussed in this section are currently being tested as nutritional supplements to prevent pathological processes involving oxygen free radicals. Although positive results have been obtained, in certain cases the protective effect of these antioxidants is low, because some of them also have a **PROOXIDANT** character. The preventive effect of antioxidants only occurs when they are administered before the oxidative lesion occurs; once the oxidative lesion has occurred, antioxidants could accelerate cell damage by reducing the metal ions released from metal proteins and thus promoting. OH formation from H_2O_2.

3. ANTIOXIDANT PLANT DEFENSE SYSTEMS

The life on earth may have been influenced by the presence of reactive oxygen species - ROS, long before the presence of an oxygenated atmosphere. Plants have a sessile life habit, and their life development is to some extent indeterminate. The metabolic pathways of plants are regulated by reactive oxygen species - ROS that act as signaling molecules in the vicinity of a stimulus, tissue, and distal organs. Due to their high reactivity and toxicity, their accumulation must be strictly controlled [16].

In plants, reactive oxygen species-ROS are mainly produced in chloroplasts and peroxisomes. However, they are also produced in the plasma membrane, cell wall, mitochondria, endoplasmic reticulum, and nucleus [17, 18].

In photosynthetic plant tissues, the chloroplast is the main site of ROS production. It is a source of $^1O_2$ no radical, the first excited electronic state of molecular oxygen O_2 and is highly reactive. The chloroplast is a sensor of environmental factors such as irradiation, heat, cold, low CO_2 levels, UV radiation, drought, salinity, and pathogens, in addition to be the organelle responsible for photosynthesis. By exposure of chloroplasts to irradiation, $^1O_2$ is constitutively generated. However, under conditions of excess light or heat shock, its production increases considerably [19, 20, 21].

In the chloroplast, in addition to producing $^1O_2$, it also produces superoxide O_2^- radicals. The superoxide radical O_2^- in the chloroplast is produced in three ways: 1) In photosystem I, where water oxidation is incomplete. It is a major reaction of photosynthesis since H_2O_2 is formed and reduced to OH or O_2^-. 2) In the b_6f complex of the electron transport chain, when plastosemiquinone PQ·- is formed from plastoquinone PQ by the reduction of plastohydroquinone-PQH$_2$. The produced plastosemiquinone interacts with O_2 to produce O_2^- [22]. 3) In photosystem I photoexcited due to electron chain overload, the electron is utilized by ferredoxin to reduce O_2 and forms O_2^-, instead of being transferred to $NADP^+$.

By *Meheler* reaction this photoreduction of O_2 takes place (reactions 72, 73, 74).

The above pathways occur under suboptimal conditions such as rapid change of light intensity, drought, heat stress, UV radiation, osmosis, and pathogen attack. However, under optimal conditions, chloroplastic $O_2^{\cdot-}$ superoxide is dismutated to H_2O_2 hydrogen peroxide by stromal superoxide dismutase enzymes and by thylakoid junctions, and then oxidized to water by peroxiredoxins in the water-water cycle and by ascorbate peroxidases [23, 24].

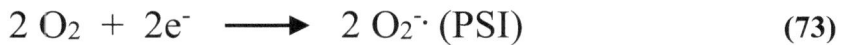

$$2\ H_2O \longrightarrow 4e^- + O_2 + 4\ H^+ \ (PSII) \tag{72}$$

$$2\ O_2 + 2e^- \longrightarrow 2\ O_2^{\cdot-} \ (PSI) \tag{73}$$

$$2\ O_2 + 2\ H^+ \longrightarrow H_2O_2 + O_2 \ (SOD) \tag{74}$$

The electrons obtained from water cleavage in photosystem II are transmitted through the electron transport chain *via* the water-water cycle. Finally, the electrons are transferred by photosystem I to O_2 for $O_2^{\cdot-}$ formation. This $O_2^{\cdot-}$ superoxide is converted to H_2O_2 by the action of the enzyme chloroplastic superoxide dismutase-SODc, while ascorbate -AsA and glutathione peroxidase - GSH convert H_2O_2 back to H_2O [25]. Therefore, the electron flux through the photosynthetic system is maintained by the water-water cycle and Asa-GSH, even when CO_2 fixation decreases. The above allows proton pumping to occur across the thylakoid membrane (Fig. **9**), thus avoiding photoinhibition by shortening the lifetime of the superoxide $O_2^{\cdot-}$ radical and decreasing the $\cdot OH$ radical.

Through photorespiration and fatty acid *β*-oxidation, ROS are produced in peroxisomes. When the plant is exposed to abiotic stress, photorespiration is an important source of reactive oxygen species-ROS [26, 27]. Glycolate oxidase is the enzyme that catalyzes glycolate oxidation to glyoxylate with H_2O_2 production, thus being the first step in the photorespiration pathway. Under light conditions in C_3 plants, H_2O_2 production in photorespiration is the main source of oxidation, considering that C_3 plants, under stress conditions or high temperatures, enhance the photorespiration process. In photosynthetic tissues, H_2O_2 peroxisomal homeostasis is carried out by catalase enzymes [28].

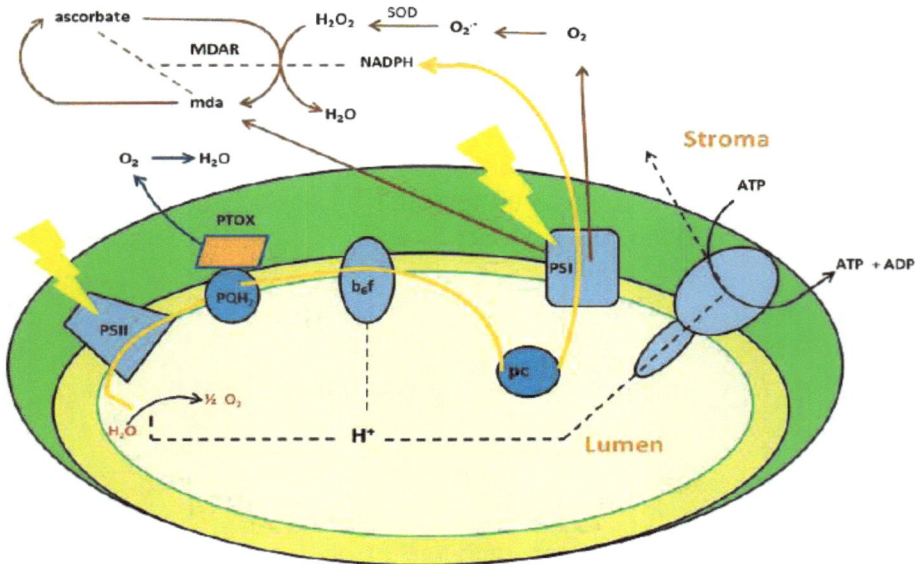

Fig. (9). Generation of ROS in chloroplasts of plant cells.

Due to electron leakage occurring in complex I, ubiquinone and complex III of the plant mitochondrial transport chain, $O_2^{\cdot-}$ superoxide is produced. This is removed by manganese superoxide dismutase -SODMn, located in the matrix. Under normal respiratory conditions, the mitochondrial electron transport chain produces a certain amount of ROS, however, under stress conditions, this amount increases considerably, which can lead to apoptosis. This occurs under abiotic stress conditions such as drought, irradiation, heavy metals or hypoxia, temperature stress, as well as under biotic stress conditions such as bacterial infection [29, 30].

The mitochondrial Alternative Oxidase 1 enzyme -AOX1, provides an electron flow that exceeds cytochrome *c* and complexes III and IV. Likewise, it catalyzes the reduction of O_2 to H_2O_2. Thus, AOX1 is the enzyme responsible for maintaining electron flow and preventing excessive reduction on the electron transport chain [31, 32]. Therefore, the rate of ROS production depends on the reduction state of electron transport chain components, a regulation that is complex due to the presence of AOX1. Mitochondrial cytochrome *c* suppression is an important event in ROS-dependent programmed cell death in plants.

NADPH oxidases present in the plasma membrane, as well as enzymes located in the plant cell wall such as: germinal oxalate oxidase, aminoxidases, lipoxygenases, class III peroxidases (PRXs) and quinone reductase, generate reactive oxygen species-ROS. Furthermore, polyamines mediate ROS signaling, participating in their production and elimination [33, 34].

Reactive oxygen species -ROS are also generated in the endoplasmic reticulum, providing an oxidative environment necessary for disulfide bridge formation in protein folding. Disulfide bridge formation in secretory proteins and in proteins associated with H_2O_2 formation is catalyzed by the enzyme oxidoreductase 1, associated with the inner membrane of the endoplasmic reticulum [35].

The intercellular space outside the plasma membrane, called the apoplast, is formed by a continuum of cell walls and extracellular spaces. Cell-cell communication is facilitated through the apoplast. The low pH of the apoplast, the presence of low amounts of glutathione and ascorbate are some of the characteristics of the apoplast that affect its redox properties. For example, its lower pH than the cytoplasm reduces the redox sensitivity of cysteine and the low presence of low molecular weight antioxidants provides less antioxidant buffering capacity. Thus, ROS levels in the apoplast lead to changes in the redox homeostasis of the cell.

Under biotic and abiotic stress conditions, apoplastic ROS are produced by ozone or pathogens. The NADPH oxidase and class III peroxidase enzymes of the cell wall catalyze ROS formation. They are soluble enzymes, are ionically or covalently bound to cell walls and their reactions are based on the peroxidative catalytic cycle and hydroxyl reactions [36]. Their reaction mechanisms are based on the hydroxyl reactions where ROS are produced with H_2O_2 reduction or $O_2^{\cdot-}$, and on the peroxidative cycle where $O_2^{\cdot-}$ is produced by H_2O_2 reduction to H_2O with various substrates oxidation. These enzymes participate in physiological processes such as lignin polymerization and during defense against pathogens.

Reactive oxygen species (ROS) have different degrees of reactivity and their half-life varies among them (Table **6**). The $^1O_2$ has a short half-life (3μs), may diffuse to 100 nm, and causes damage to nucleic acids, proteins, pigment, and lipids. $O_2^{\cdot-}$ with a half-life of 2-4 μs may migrate distances greater than 30 nm. Although its reactivity is moderate, through the *Haber-Weiss* reaction it intervenes in ·OH production, a much more reactive and toxic species. H_2O_2 has a longer half-life of 1 ms, may migrate to distances greater than 1 μm, has moderate reactivity and has the capacity to cross membranes by means of aquaporins [37, 38, 39].

Table 6. Half-life of reactive oxygen species -ROS in plants.

ROS	$t_{1/2}$	Migration distance	Production site	Trappping systems
Singlet oxygen ($^1O_2$)	1-4 µs	30nm	Membranes, chloroplasts, nuclei	Carotenoids and α-tocopherol
Hydrogen peroxide (H_2O_2)	>1ms	>1µm	Peroxisomes, chloroplasts, mitochondria, cytosol, apoplast	APX, CAT, GPX, PER, PRX, ascorbate, gluthation
Superoxide ($O_2^{\cdot -}$)	1-4µs	30nm	Apoplast, chloroplasts, mitochondria, peroxisomes, electron transfer chain.	SOD, flavonoids, ascorbate.
Hydroxyl radical (·OH)	1ns	1nm	Iron and H_2O_2 (Fenton reaction)	Flavonoids, proline, sugars, ascorbate.

In turn, the ·OH hydroxyl radical has a half-life of 1µs and is the most reactive radical. It may migrate to 1 nm and causes damage to nucleic acids, proteins, and lipids by interacting with cellular molecules. Excess ·OH radicals accumulate and cause apoptosis, when there is not an enzymatic system capable of eliminating them.

The enzymatic antioxidants include catalases (CATs), superoxide dismutases (SODs), glutathione reductase (GRs), glutathione transferases (GSTs), glutathione peroxidase (GPXs), ascorbate peroxidases (APXs), dehydroascorbate reductases (DHARs) and monodehydroascorbate reductase (MDHARs).

Metalloenzymes include the enzyme superoxide dismutase-SOD. SODs is found in chloroplasts Fe-SOD and Cu/Zn-SOD, mitochondria SOD-Mn, cytosol Cu/Zn-SOD and peroxisomes Cu/Zn-SOD. They catalyze $O_2^{\cdot -}$ superoxide removal, through O_2 and H_2O_2 dismutation, decreasing ·OH formation through the *Haber-Weiss* reaction. As a response of the plant to biotic and abiotic stress, high production and activity of SOD occurs. In addition, the plant is protected from damage caused by oxidative stress through overexpression of SOD [40, 41].

In plant cells H_2O_2 is degraded by enzymes such as catalases. These -CATs enzymes possess a *heme* group and catalyze H_2O_2 dismutation into H_2O and O_2, have different specificity and subcellular localization. The main H_2O_2 producers in plant cells are peroxisomes; however, CATs are also found in the cytosol, mitochondria, and chloroplasts. Biotic and abiotic stress conditions such as bacterial infection, nematodes, salinity, drought, UV radiation, heat and cold induce the expression and activity of CATs [42].

In the cytosol, mitochondria, chloroplasts, and peroxisomes, H_2O_2 is removed by the arcorbate-glutamate -AsA-GSH cycle or Foyer-Halliwell-Asada cycle, which is under the control of the enzymes glutathione reductase (GR), ascorbate peroxidase (APX), dehydroascorbate reductase (DHAR) and monode hydroascorbate reductase (MDHAR).

The glutathione peroxidase enzymes GPXs utilize GSH to remove H_2O_2 and hydroperoxides by maintaining H_2O_2 homeostasis. They are found in substrate specificity, in the cytosol, mitochondria, chloroplast and endoplasmic reticulum. By maintaining the $NADPH/NADP^+$ or thiol/disulfide balance, they are involved in redox homeostasis regulation and constitute a connection between the glutathione-thioredoxin systems. Environmental conditions such as cold, drought, metal treatment, oxidative stress and pathogens induce the production and activity of GPXs [43].

The glutathione transferase enzymes -GSTs constitute a large group of enzymes that catalyze the conjugation of GSH to electrophilic sites on peroxidized lipids or phytotoxins. They are found in nucleus, cytosol, chloroplast, mitochondrion and apoplast. In addition, participate in growth and developmental processes, as well as in response to biotic and abiotic factors [44].

Additionally, to the above enzymatic antioxidants low molecular weight compounds such as GSH, AsA, α-tocopherol, carotenoids, proline, and flavonoids [45] exert their ROS scavenging activity as non-enzymatic antioxidants in plants.

In plants, ascorbic acid - AsA is the most abundant antioxidant. It eliminates ROS directly or through the AsaA-GSH cycle. GSH is a reductant of ROS and a substrate for certain peroxidases. GSH is found in most cellular organelles such as the nucleus, cytosol, chloroplast, mitochondria, endoplasmic reticulum, peroxisomes, vacuole and apoplast. The *Cys* thiol group of GSH in the reduced state, through thiol-disulfide exchange, donates a reducing equivalent to proteins and ROS molecules. After hydrogen donation the radical GSH reacts with another GSH radical to form the glutathione disulfide GSSH. An increase in GSH confers tolerance to salinity, drought, heavy metals, and high/low temperatures [46].

Tocopherols are effective lipophilic antioxidants that protect biological membranes by scavenging ROS and lipid radicals. They are $^1O_2$ scavengers, prevent lipid peroxidation and protect photosystem II. Among the four isomers present in plants, α-tocopherol (vitamin E) has the highest antioxidant capacity. In response to biotic and abiotic stress the cellular levels of α-tocopherol vary mainly as a response to osmotic stress and irradiation [47, 48].

Another group of lipid antioxidant present in photosynthetic and non-photosynthetic plant organelles are carotenoids. They can absorb light and transfer the energy to chlorophyll molecules by forming antennae on the photosystems.

Through the xanthophyll cycle, they dissipate excess activation energy thus preventing the formation of $^1O_2$ by the activated chlorophyll reaction. Tolerance to abiotic stress improves with increasing carotenoid levels in the plant cell [49].

Proline is a potent antioxidant that scavenges free radicals and inhibits lipid peroxidation and apoptosis. It possesses a compartmentalized metabolism that allows it to be transported between the cytosol, chloroplast, and mitochondria. Tolerance to abiotic stresses such as salinity, drought and low temperatures is mainly due to high levels of proline [50].

CONCLUSION

An analysis of reaction mechanisms and kinetic behavior of the antioxidant enzymes that form part of the defense system developed by cells to avoid the adverse effects of the accumulation of reactive oxygen species -ROS has been performed. This defense system also includes other primary and secondary defense mechanisms that include besides antioxidant enzymes, mitochondrial defense system, proteins, vitamins, carotenoids, and the antioxidant defense system that plants have developed to avoid the harmful effects of reactive oxygen species.

REFERENCES

[1]   Bafana, A.; Dutt, S.; Kumar, A.; Kumar, S.; Ahuja, P.S. The basic and applied aspects of superoxide dismutase. *J. Mol. Catal., B Enzym.,* **2011**, *68*, 129-138.
        [http://dx.doi.org/10.1016/j.molcatb.2010.11.007]

[2]   Xu, K.Y.; Kuppusamy, P. Dual effects of copper-zinc superoxide dismutase. *Biochem. Biophys. Res. Commun.,* **2005**, *336*(4), 1190-1193.
        [http://dx.doi.org/10.1016/j.bbrc.2005.08.249] [PMID: 16169521]

[3]   Liochev, S.I.; Fridovich, I. The effects of superoxide dismutase on H_2O_2 formation. *Free Radic. Biol. Med.,* **2007**, *42*(10), 1465-1469.
        [http://dx.doi.org/10.1016/j.freeradbiomed.2007.02.015] [PMID: 17448892]

[4]   Liochev, S.I.; Fridovich, I. Mechanism of the peroxidase activity of Cu, Zn superoxide dismutase. *Free Radic. Biol. Med.,* **2010**, *48*(12), 1565-1569.
        [http://dx.doi.org/10.1016/j.freeradbiomed.2010.02.036] [PMID: 20211248]

[5]   Bonini, M.G.; Gabel, S.A.; Ranguelova, K.; Stadler, K.; Derose, E.F.; London, R.E.; Mason, R.P. Direct magnetic resonance evidence for peroxymonocarbonate involvement in the cu,zn-superoxide dismutase peroxidase catalytic cycle. *J. Biol. Chem.,* **2009**, *284*(21), 14618-14627.
        [http://dx.doi.org/10.1074/jbc.M804644200] [PMID: 19286663]

[6]   Medinas, D.B.; Toledo, J.C., Jr; Cerchiaro, G.; do-Amaral, A.T.; de-Rezende, L.; Malvezzi, A.; Augusto, O. Peroxymonocarbonate and carbonate radical displace the hydroxyl-like oxidant in the Sod1 peroxidase activity under physiological conditions. *Chem. Res. Toxicol.,* **2009**, *22*(4), 639-648.
        [http://dx.doi.org/10.1021/tx800287m] [PMID: 19243126]

[7]   Escobar, J.A.; Rubio, M.A.; Lissi, E.A. Sod and catalase inactivation by singlet oxygen and peroxyl radicals. *Free Radic. Biol. Med.,* **1996**, *20*(3), 285-290.
        [http://dx.doi.org/10.1016/0891-5849(95)02037-3] [PMID: 8720898]

[8]   Martínez, C.M. Estrés oxidativo y mecanismo de defensa antioxidante. Cap. 18.*GIL H. A. Tratado de nutrición*; Tomo I. Editorial Panamericana, **2010**, pp. 455-480.

[9]   Echtay, K.S. Mitochondrial uncoupling proteins--what is their physiological role? *Free Radic. Biol. Med.,* **2007**, *43*(10), 1351-1371.
[http://dx.doi.org/10.1016/j.freeradbiomed.2007.08.011] [PMID: 17936181]

[10]   Brand, M.D.; Affourtit, C.; Esteves, T.C.; Green, K.; Lambert, A.J.; Miwa, S.; Pakay, J.L.; Parker, N. Mitochondrial superoxide: production, biological effects, and activation of uncoupling proteins. *Free Radic. Biol. Med.,* **2004**, *37*(6), 755-767.
[http://dx.doi.org/10.1016/j.freeradbiomed.2004.05.034] [PMID: 15304252]

[11]   James, A.M.; Smith. R.A.J.; Murphy, M.P. Antioxidant and prooxidant properties of mitochondrial Coenzyme Q. *Arch. Biochem. Biophys.,* **2004**, *423*(1), 47-56.
[http://dx.doi.org/10.1016/j.abb.2003.12.025] [PMID: 14989264]

[12]   Maroz, A.; Anderson, R.F.; Smith, R.A.J.; Murphy, M.P. Reactivity of ubiquinone and ubiquinol with superoxide and the hydroperoxyl radical: implications for *in vivo* antioxidant activity. *Free Radic. Biol. Med.,* **2009**, *46*(1), 105-109.
[http://dx.doi.org/10.1016/j.freeradbiomed.2008.09.033] [PMID: 18977291]

[13]   Sies, H. Glutathione and its role in cellular functions. *Free Radic. Biol. Med.,* **1999**, *27*(9-10), 916-921.
[http://dx.doi.org/10.1016/S0891-5849(99)00177-X] [PMID: 10569624]

[14]   Yamahuchi, R.; Vitamin, E. Mechanism of its antioxidant activity. *Food Sci. Technol. Int.,* **1997**, *3*, 301-309.

[15]   Roberts, L.J., II; Oates, J.A.; Linton, M.F.; Fazio, S.; Meador, B.P.; Gross, M.D.; Shyr, Y.; Morrow, J.D. The relationship between dose of vitamin E and suppression of oxidative stress in humans. *Free Radic. Biol. Med.,* **2007**, *43*(10), 1388-1393.
[http://dx.doi.org/10 1016/j.freeradbiomed.2007.06.019] [PMID: 17936185]

[16]   Czarnocka, W.; Karpiński, S. Friend or foe? Reactive oxygen species production, scavenging and signaling in plant response to environmental stresses. *Free Radic. Biol. Med.,* **2018**, *122*, 4-20.
[http://dx.doi.org/10.1016/j.freeradbiomed.2018.01.011] [PMID: 29331649]

[17]   Das, K.; Roychoudhury, A. Reactive oxygen species (ROS) and response of antioxidants as ROS-scavengers during environmental stress in plants. *Front. Environ. Sci.,* **2014**, 2.
[http://dx.doi.org/10.3389/fenvs.2014.00053]

[18]   Karpinska, B.; Alomrani, S.O.; Foyer, C.H. Inhibitor-induced oxidation of the nucleus and cytosol in Arabidopsis thaliana: implications for organelle to nucleus retrograde signaling. *Philosophical Transactions B.,* **2017**, *372*

[19]   Foyer, C.H.; Noctor, G. Redox regulation in photosynthetic organisms: signaling, acclimation, and practical implications. *Antioxid. Redox Signal.,* **2009**, *11*(4), 861-905.
[http://dx.doi.org/10.1089/ars.2008.2177] [PMID: 19239350]

[20]   Mullineaux, P.; Karpinski, S. Signal transduction in response to excess light: getting out of the chloroplast. *Curr. Opin. Plant Biol.,* **2002**, *5*(1), 43-48.
[http://dx.doi.org/10.1016/S1369-5266(01)00226-6] [PMID: 11788307]

[21]   Wituszyńska, W.; Karpiński, S. Programmed Cell Death as a Response to High Light, UV, and Drought Stress in Plants.

[22]   Pospíšil, P. Production of Reactive Oxygen Species by Photosystem II as a Response to Light and Temperature Stress. *Front. Plant Sci.,* **2016**, *7*, 1950.
[http://dx.doi.org/10.3389/fpls.2016.01950] [PMID: 28082998]

[23]   Mignolet-Spruyt, L.; Xu, E.; Idänheimo, N.; Hoeberichts, F.A.; Mühlenbock, P.; Brosché, M.; Van Breusegem, F.; Kangasjärvi, J. Spreading the news: subcellular and organellar reactive oxygen species

production and signalling. *J. Exp. Bot.,* **2016,** *67*(13), 3831-3844.
[http://dx.doi.org/10.1093/jxb/erw080] [PMID: 26976816]

[24]   Mubarakshina, M.M.; Ivanov, B.N. The production and scavenging of reactive oxygen species in the plastoquinone pool of chloroplast thylakoid membranes. *Physiol. Plant.,* **2010,** *140*(2), 103-110.
[http://dx.doi.org/10.1111/j.1399-3054.2010.01391.x] [PMID: 20553418]

[25]   Foyer, C.H.; Halliwell, B. The presence of glutathione and glutathione reductase in chloroplasts: A proposed role in ascorbic acid metabolism. *Planta,* **1976,** *133*(1), 21-25.
[http://dx.doi.org/10.1007/BF00386001] [PMID: 24425174]

[26]   Weber, A.P.M.; Bauwe, H. Photorespiration--a driver for evolutionary innovations and key to better crops. *Plant Biol.,* **2013,** *15*(4), 621-623.
[http://dx.doi.org/10.1111/plb.12036] [PMID: 23786418]

[27]   Maurino, V.G.; Peterhansel, C. Photorespiration: current status and approaches for metabolic engineering. *Curr. Opin. Plant Biol.,* **2010,** *13*(3), 249-256.
[http://dx.doi.org/10.1016/j.pbi.2010.01.006] [PMID: 20185358]

[28]   Mhamdi, A.; Noctor, G.; Baker, A. Plant catalases: peroxisomal redox guardians. *Arch. Biochem. Biophys.,* **2012,** *525*(2), 181-194.
[http://dx.doi.org/10.1016/j.abb.2012.04.015] [PMID: 22546508]

[29]   Gupta, K.J.; Igamberdiev, A.U. Reactive nitrogen species in mitochondria and their implications in plant energy status and hypoxic stress tolerance. *Front. Plant Sci.,* **2016,** *7*, 369.
[http://dx.doi.org/10.3389/fpls.2016.00369] [PMID: 27047533]

[30]   Petrov, V.; Hille, J.; Mueller-Roeber, B.; Gechev, T.S. ROS-mediated abiotic stress-induced programmed cell death in plants. *Front. Plant Sci.,* **2015,** *6*, 69.
[http://dx.doi.org/10.3389/fpls.2015.00069] [PMID: 25741354]

[31]   Cvetkovska, M.; Alber, N.A.; Vanlerberghe, G.C. The signaling role of a mitochondrial superoxide burst during stress. *Plant Signal. Behav.,* **2013,** *8*(1)e22749
[http://dx.doi.org/10.4161/psb.22749] [PMID: 23221746]

[32]   Navrot, N.; Rouhier, N.; Gelhaye, E.; Jacquot, J-P. Reactive oxygen species generation and antioxidant systems in plant mitochondria. *Physiol. Plant.,* **2007,** *129*, 185-195.
[http://dx.doi.org/10.1111/j.1399-3054.2006.00777.x]

[33]   Camejo, D.; Guzmán-Cedeño, Á.; Moreno, A. Reactive oxygen species, essential molecules, during plant-pathogen interactions. *PPB,* **2016,** *103*, 10-23.
[http://dx.doi.org/10.1016/j.plaphy.2016.02.035] [PMID: 26950921]

[34]   Pottosin, I.; Velarde-Buendía, A.M.; Bose, J.; Zepeda-Jazo, I.; Shabala, S.; Dobrovinskaya, O. Cross-talk between reactive oxygen species and polyamines in regulation of ion transport across the plasma membrane: implications for plant adaptive responses. *J. Exp. Bot.,* **2014,** *65*(5), 1271-1283.
[http://dx.doi.org/10.1093/jxb/ert423] [PMID: 24465010]

[35]   Ozgur, R.; Turkan, I.; Uzilday, B.; Sekmen, A.H. Endoplasmic reticulum stress triggers ROS signalling, changes the redox state, and regulates the antioxidant defence of *Arabidopsis thaliana*. *J. Exp. Bot.,* **2014,** *65*(5), 1377-1390.
[http://dx.doi.org/10.1093/jxb/eru034] [PMID: 24558072]

[36]   Passardi, F.; Penel, C.; Dunand, C. Performing the paradoxical: how plant peroxidases modify the cell wall. *Trends Plant Sci.,* **2004,** *9*(11), 534-540.
[http://dx.doi.org/10.1016/j.tplants.2004.09.002] [PMID: 15501178]

[37]   Bienert, G.P.; Møller, A.L.B.; Kristiansen, K.A.; Schulz, A.; Møller, I.M.; Schjoerring, J.K.; Jahn, T.P. Specific aquaporins facilitate the diffusion of hydrogen peroxide across membranes. *J. Biol. Chem.,* **2007,** *282*(2), 1183-1192.
[http://dx.doi.org/10.1074/jbc.M603761200] [PMID: 17105724]

[38]   Halliwell, B. Reactive species and antioxidants. Redox biology is a fundamental theme of aerobic life.

Plant Physiol., **2006**, *141*(2), 312-322.
[http://dx.doi.org/10.1104/pp.106.077073] [PMID: 16760481]

[39]   Hatz, S.; Lambert, J.D.C.; Ogilby, P.R. Measuring the lifetime of singlet oxygen in a single cell: addressing the issue of cell viability. *Photochem. Photobiol. Sci.,* **2007**, *6*(10), 1106-1116.
[http://dx.doi.org/10.1039/b707313e] [PMID: 17914485]

[40]   Gupta, A.S.; Webb, R.P.; Holaday, A.S.; Allen, R.D. Overexpression of Superoxide Dismutase Protects Plants from Oxidative Stress (Induction of Ascorbate Peroxidase in Superoxide Dismutase-Overexpressing Plants). *Plant Physiol.,* **1993**, *103*(4), 1067-1073.
[http://dx.doi.org/10.1104/pp.103.4.1067] [PMID: 12232001]

[41]   Mittler, R. Oxidative stress, antioxidants and stress tolerance. *Trends Plant Sci.,* **2002**, *7*(9), 405-410.
[http://dx.doi.org/10.1016/S1360-1385(02)02312-9] [PMID: 12234732]

[42]   Caverzan, A.; Casassola, A.; Brammer, S.P. Antioxidant responses of wheat plants under stress. *Genet. Mol. Biol.,* **2016**, *39*(1), 1-6.
[http://dx.doi.org/10.1590/1678-4685-GMB-2015-0109] [PMID: 27007891]

[43]   Bela, K.; Horváth, E.; Gallé, Á.; Szabados, L.; Tari, I.; Csiszár, J. Plant glutathione peroxidases: emerging role of the antioxidant enzymes in plant development and stress responses. *J. Plant Physiol.,* **2015**, *176*, 192-201.
[http://dx.doi.org/10.1016/j.jplph.2014.12.014] [PMID: 25638402]

[44]   Dixon, D.P.; Skipsey, M.; Edwards, R. Roles for glutathione transferases in plant secondary metabolism. *Phytochemistry,* **2010**, *71*(4), 338-350.
[http://dx.doi.org/10.1016/j.phytochem.2009.12.012] [PMID: 20079507]

[45]   Petrussa, E.; Braidot, E.; Zancani, M.; Peresson, C.; Bertolini, A.; Patui, S.; Vianello, A. Plant flavonoids--biosynthesis, transport and involvement in stress responses. *Int. J. Mol. Sci.,* **2013**, *14*(7), 14950-14973.
[http://dx.doi.org/10.3390/ijms140714950] [PMID: 23867610]

[46]   Hasanuzzaman, M.; Nahar, K.; Anee, T.I.; Fujita, M. Glutathione in plants: biosynthesis and physiological role in environmental stress tolerance. *Physiol. Mol. Biol. Plants,* **2017**, *23*(2), 249-268.
[http://dx.doi.org/10.1007/s12298-017-0422-2] [PMID: 28461715]

[47]   Havaux, M.; Eymery, F.; Porfirova, S.; Rey, P.; Dörmann, P.; Vitamin, E. Vitamin E protects against photoinhibition and photooxidative stress in Arabidopsis thaliana. *Plant Cell,* **2005**, *17*(12), 3451-3469.
[http://dx.doi.org/10.1105/tpc.105.037036] [PMID: 16258032]

[48]   Kamal-Eldin, A.; Appelqvist, L.A. The chemistry and antioxidant properties of tocopherols and tocotrienols. *Lipids* **1996**, *31*(7), 671-701.
[http://dx.doi.org/10.1007/BF02522884] [PMID: 8827691]

[49]   Nisar, N.; Li, L.; Lu, S.; Khin, N.C.; Pogson, B.J. Carotenoid metabolism in plants. *Mol. Plant,* **2015**, *8*(1), 68-82.
[http://dx.doi.org/10.1016/j.molp.2014.12.007] [PMID: 25578273]

[50]   Kaur, G.; Asthir, B. Proline: a key player in plant abiotic stress tolerance. *Biol. Plant.,* **2015**, *59*, 609-619.
[http://dx.doi.org/10.1007/s10535-015-0549-3]

CHAPTER 5

Flavonoids as Reactive Oxygen Species Promotors

Abstract: The high enzymatic reactivity of reactive oxygen species can be seen in the metabolism of flavonoids, since throughout the metabolic process that presents this type of chemical compounds as trapping agents of reactive oxygen species or antioxidants, finally and due to the speed of reaction they themselves become promoting agents of the same reactive oxygen species. The flavonoids are organic molecules that, due to their chemical nature and their low redox potential ($0.23 < E7 < 0.75$ V), can easily react with oxygen-free radicals, inhibiting both the action of the radicals and the molecules producing them; through antioxidant action mechanisms; some of which are explained by physicochemical and molecular parameters such as heat formation (ΔH_f), Ionization potential and (IP Bond dissociation energy (BDE). However, after bioavailability and absorption, the flavonoids promote oxygen free radical's production acting as prooxidants mainly through hydrogen atom transfer HAT or simple electron transfer-SET mechanisms, properties and mechanisms of antioxidant and prooxidant reaction.

Keywords: Antioxidants, Bioavailability, Bond Dissociation Energy – BDE, Flavonoids, Formation heat of flavonoids radical -ΔH_f, Hydrogen atom transfer - HAT, Measurement -ROS, Prooxidant enzymes mechanism, Pro-oxidant mechanisms of flavonoids, Quantitative Structure-Activity Relationship-QSAR, Radical trapping, Redox chemistry, Single electron transfer -SET, Structure-Activity Relationship-SAR.

INTRODUCTION

Polyphenols are compounds produced by plants as a product of their secondary metabolism and represent the most abundant and widely distributed groups of substances in plants. Polyphenols are biosynthesized by two main pathways: the shikimic acid pathway and the malonic acid pathway [1]. They are involved in plant reproduction, growth, mechanical support, and resistance to pathogens and predators, harvest protection and seed storage, absorption of prejudicial ultraviolet radiation and allelopathic mechanisms [2], among others.

1. FLAVONOID CHEMICAL STRUCTURE

Polyphenols can be classified into phenolic acids, flavonoids, stilbenes and lignans, considering the number of phenolic rings they contain, and the substituent groups present on these rings. Flavonoids are one of the main groups of phenolic compounds. The chemical nature of flavonoids depends on their structure, degree of hydroxylation, substitutions, conjugations, and degree of polymerization.

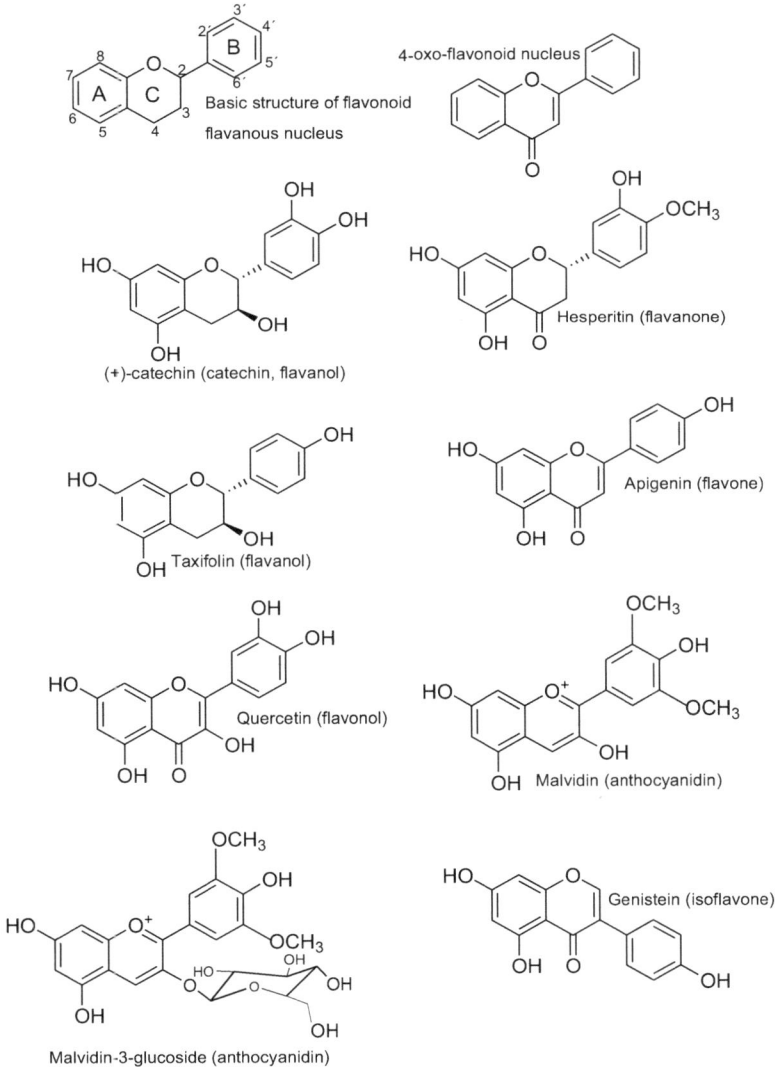

Fig. (1). Flavonoids. Basic structure and types.

They are constituted structurally by a basic skeleton of 15 carbons organized in two aromatic rings (A and B) connected by three carbon atoms that form an oxygenated heterocycle (C ring). The A ring, originated by condensation of three units of acetate pathway malonic acid and the B ring together with the three carbon atoms, constitute a unit of phenylpropanoid, biosynthesized by the pathway of shikimic acid.

Theoretically it is possible to obtain a very large number of different flavonoid structures if 10 carbons of the basic flavonoid backbone are substituted by different groups such as methyl, methoxyl, hydroxyl, isoprenyl and benzyl. Additionally, each hydroxyl group and some carbons can be substituted by one or more sugars and in turn each sugar can be alcyled by several phenolic or aliphatic acids. According to Williams & Grayer, 2004 over 9000 different flavonoids of plant origin have been reported [3], [4].

Generally, the classification of flavonoids is based on the degree of oxidation of the three-carbon bridge and on the presence and way the hydroxyl and methyl groups are attached to the basic molecule. Depending on their structural characteristics (Fig. **1**) flavonoids can be classified into:

1. Flavanols as catechin, with a group–OH in position 3 of ring C.
2. Flavonols, represented by quercetin possessing a carbonyl group in position 4 and a–OH group in position 3 of the C ring.
3. Anthocyanidins have the–OH group attached to the 3 position and in the C ring a double bond between carbons 3 and 4.
4. Flavones, such as diosmethine, have a carbonyl group at position 4 of the C-ring and at position C3 it lacks the hydroxyl group.
5. Flavanones, they lack the hydroxyl group in position C3.
6. Isoflavonoids vary from the previous ones since the B ring is attached to the C3 of the C ring instead of the C2 ring.

In plants flavonoids are present mainly as glycosides. Flavonols and flavones are bound to sugars, preferably to the C3 position and less frequently to the C7 of ring A, so that these compounds are generally found as *O*-glycosides, where the most frequent sugar residue is D-glucose. There are also sugar residues such as D-galactose, L-arabinose, L-rhamnose, D-xylose and D-glucuronic acid. The structure of the flavonoid molecule without sugar is called an aglycone. Glycosides are more soluble in water and less reactive in respect to free radicals than their aglycone or respective flavonoid [5].

2. TYPES AND SOURCES OF FLAVONOIDS

Flavonoids are found in edible fruits and vegetables; however, they may be present in one or more foods [6] and the type of flavonoid varies according to source [7].

Flavonoids are found in flowers, fruits, seeds, and vegetables, as well as in green tea, black tea, wine, beer and soybeans, which are consumed regularly in the human diet and can also be used in the form of nutritional supplements, along with certain vitamins and minerals.

The flavonols group includes quercetin, kaempferol and myricetin found mainly in onions, broccoli, apples, cherries, sorrel, berries, and tea [8], among others. The most common flavonol in the diet is quercetin. Quercetin is a yellow-green flavonoid found in cherries, grapes, apples, onions, broccoli, and red cabbage. Although quercetin is found in vegetable foods at low levels (10 mg/kg), in onions it is found in high concentrations (> 1.2 g/kg fresh weight). Tea contains considerable amounts of quercetin and kaempferol and appears in different glycosylated forms. In black tea, quercetin levels are between 10 and 25 mg/L, while myricetin and kaempferol range between 2 to 5 mg/L and 7 to 17 mg/L, respectively.

Catechins, epicatechins and galocatechins belong to the flavanols. Although many fruits and vegetables contain catechins, their levels range from 4.5 mg/kg in kiwi, 250 mg/kg in apricot to 610 mg/kg in dark chocolate [9]. They are also found in tea (green tea contains 200 - 800 mg/L of catechin) [10], which also contains a high level of phenolic compounds > 35% of dry weight. Catechins are usually presented as esterified with gallic acid o aglycones. Anthocyanins are responsible for the blue, red, or violet colors of vegetables and fruits such as aubergines, apples, rhubarb, black grapes, and some berries. They include pelargonidine, malvidine and cyanidine [11]. The most common anthocyanin in food is cyanidine. The content is proportional to the intensity of the color and reaches values higher than 2-4 g/kg fresh weight in black berries.

Flavanones include hesperidin, rutin, naranjin and eriodictyol. Hesperidin is found in citrus fruits such as mandarin and orange in both solid tissue and juice and eriodictol is found in lemons. Also found in plums and tomatoes. The hesperidin content in orange juice is around 200 to 600 mg/L and about 40 to 140 mg of flavanone glycosides are found in a glass of orange juice.

Flavones such as luteolin, apigenin and diosmetin can be found in parsley, thyme, celery, and red bell pepper. However, citrus skin contains considerable amounts of polymethoxy flavones: tangerine, nobiletine and synensetine (> 6.5 g/L mandarin essential oil).

In legumes and soybeans are found mainly genistein and daidzein belonging to isoflavonoids, which are flavonoids with estrogen-like structure [12]. The isoflavonoid content of soybeans is between 580 and 3800 mg isoflavonoids/kg fresh weight and soy milk contains between 30 and 175 mg/L [13]. Isoflavonoids have also been found in families other than legumes [14].

The taste of the wine is influenced by the phenolic components, especially the flavonoids. Approximately 200 equiv/L is the total phenolic content; red wine contains about 120 mg/L of anthocyanins, 750 mg/L of anthocyanogenic tannins, 250 mg/L of catechins, and 50 mg/L of flavonols.

Although food habits are very diverse in the world, the average value of flavonoid intake is estimated at 23mg/day, with flavonols being predominant, especially quercetin. The main food sources of flavonols include black tea, onions, apples, black pepper, which contains about 4g/kg quercetin, and alcoholic beverages such as wine and beer.

Flavonoids are studied for their structural diversity, biological and ecological [15], importance and for their properties as health promoting agents [16], [11]. Numerous research have mainly studied the activity of flavonoids as antioxidants [17], in coronary diseases [18], as anti-inflammatories [19], in vascular activity [20], as anticancer agents [21], antimicrobials [22] and as neuroprotectors, among others [23]. In a review performed by Lago *et al.*, 2014 [24], the anti-inflammatory effect of flavonoids utilized in the treatment of lung diseases was demonstrated. In a study realized by Mattioli *et al.*, 2020 [25] evaluated the effect of different types of flavonoids consumption on the incidence of chronic respiratory disease in adults and found that flavanones may reduce the risk of non-allergic rhinitis.

These functions of flavonoids are related to their chemical structure. Jeong *et al.*, 2007 [26] studies concluded that flavonoids with the 3–OH group have antioxidant activity, flavonoids with the 5–OH and/or 7–OH groups present high cytotoxicity and flavonoids with the 3'–OMe and/or 5'–OMe group present P-protein inhibitory effect, while flavonoids with the 6–OMe group do not have it.

In plants, chalcones are the precursors of flavonoids and are synthesized from glucose *via* shikimic acid and HAc-malonic acid, respectively. Given the importance of these compounds as precursors of flavonoids, numerous studies have been carried out to obtain them through organic synthesis. A detailed analysis of the methods employed for organic chalcone synthesis and new pathways for synthesis of compounds as flavonol-3-ols are presented by Espíndola, 2020 [27].

3. FLAVONOIDS AS TRAPPING AGENTS FOR REACTIVE OXYGEN SPECIES -ROS.

The intake of chemical compounds such as flavonoids is one of the pathways that may reduce reactive oxygen species (ROS) levels in the body. Flavonoids, due to their low redox potential ($0.23 < E7 < 0.75$ V), may reduce free radicals that are highly oxidized with redox potential values from $2.13 - 1.0$ V (O^{-}_{2}, $\cdot OH$, $NO^{\cdot}$, $RO^{\cdot}$, $ROO^{\cdot}$). In addition, flavonoids can suppress the ROS production due to the redox enzyme's inhibition such as (monooxigenases, cyclooxigenases, lipooxigenases, xanthine oxidase and NAD(P)H oxidases) and bind to transition metal ions which are involved in the oxygen radicals generation *via* Fenton reaction. However, flavonoids in addition to their activity as antioxidants also have prooxidant activity.

3.1. Structure-function Relationship

Structure-activity relationship (SAR) and quantitative structure-activity relationship (QSAR) are useful tools to determine the antioxidant nature of flavonoids by correlating physicochemical or molecular descriptors of structurally related compounds with their biological activity or physical properties. Parameters such as electronic properties, hydrophobicity, topology, and steric effects characterize the molecular descriptors.

3.1.1. Flavonoids as Antioxidants

The main activity of flavonoids is their antioxidant capacity. The antioxidant effect of flavonoids is attributed to the presence of phenolic hydroxyl groups attached to the ring structure. They exert this effect as reactive oxygen species trapping agents, singlet oxygen traps, hydrogen donor compounds and metal ion chelators.

3.1.1.1. Structural Determinants for Antioxidant Activity of Flavonoids

A flavonoid's chemical structure directly influences its antioxidant capacity. Thus, the number and position of the hydroxyl groups in the B and A rings and the degree of conjugation between the B and C rings determine the level of the antioxidant effect of the flavonoid. Structural determinants exist to determine the level of radical trapping and/or antioxidant potential of flavonoids and they are known as the Bors' criteria:

1. The *o*-dihydroxy (3',4'−diOH) structure in the B-ring gives stability to the flavonoid phenoxyl radical through hydrogen bonding or by electronic delocalization.
2. Conjugation of the C2−C3 double bond with the 4-*oxo* group determines the coplanarity of the hetero-ring.
3. The presence of the 3−OH and 5−OH groups contribute to the maximum radical trapping capacity.
4. In the absence of the *o*-hydroxy structure on the B-ring, hydroxyl substituents on the A-ring in a catechol structure can both compensate for and be a determinant of antiradical activity.

The structural criteria that determine the antioxidant activity of flavonoids are shown in Figure. (2) [28].

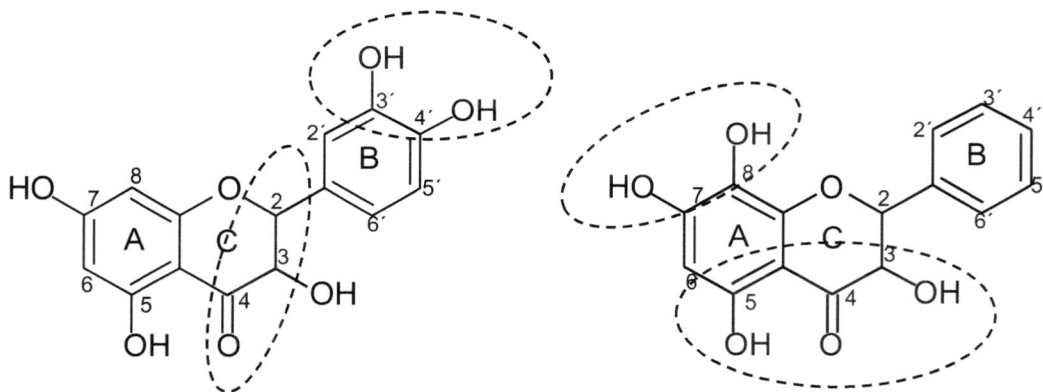

Fig. (2). Structural criteria for antioxidant activity of flavonoids.

The effect of flavonoids as antioxidants on the inhibition of free radicals at the cellular level involves mainly their action as trapping agents of superoxide $O_2^{\cdot-}$ radicals; on lipid peroxidation by their reaction with peroxyl radicals and on the formation of ·OH by iron ion chelation.

Sichel *et al.* in 1991 [29] evaluated the ability of flavonoids as superoxide ($O_2^{\cdot-}$) radical trapping agents using the spectrometric technique: Electronic Spin Resonance - ESR. For it, they generated *in vitro* alkaline solutions that contained the superoxide anion $O_2^{\cdot-}$ and concluded that $O_2^{\cdot-}$ generation mechanisms in alkaline solutions can involve: 1) molecular oxygen reduction by CH_3COCH_2- or CH_3SOCH_2- which is produced by hydrogen peroxide dismutation, 2) hydrogen peroxide reaction with deprotoned hydrogen peroxide in which the $\cdot OH$ and $\cdot HO_2$ radicals are important. All evaluated flavonoids trapped the O^-_2 radical with some relationship between concentration and trapping efficiency. The most effective flavonoids were in the following order: pelargonidine > quercetin > oenine > quercethrin > catechin > malvine > cyanidine > pelargonine > routine > calistefine > feonidine > morine > apigenin.

The trapping capacity of these compounds is related to the property of forming stable radicals. By ESR spectroscopy it has been determined that aromatic compounds such as flavonoids with hydroxyl groups and mainly with an *ortho*-dihydroxy function in the B-ring or carbonyl near the 3−OH ring, can originate stable radicals, as in the case of quercetin or alkyl gallates.

The introduction of the OH group at position 3 called flavonol-3-ol, as in the case of catechin, also increases the trapping capacity, because the C−ring only acquires certain orientations due to the existence at position 3 the OH group [30].

The aglicones of the flavon-3-ols present high activity; this suggests that the chemical mechanisms of the trapping capacity also involve the carbonyl and OH groups in positions 4 and 3, respectively. It has been observed that the presence of two hydroxyl groups in the B-ring gives catechin a high entrapment capacity, while apigenin with only one hydroxyl in the B-ring has a moderate entrapment capacity.

Potapovich and Kostyuk in 2003 [31], evaluated the antioxidant capacity of several flavonoids with similar chemical structure and found that the high antiradical capacity of flavonoids on superoxide O^-_2, is due to the reactivity of the hydroxyl groups in the *meta*-position of ring A and in the *ortho*-position of ring B. And as mentioned in the previous paragraph, the antioxidant activity of flavonoids depended on the presence of hydroxyl groups in their structure. The high activity found for EGCG and ECG catechins is due to the addition of a gallic acid residue in C−3; on the contrary, because glycosylation chemically blocks active groups of quercetin, its rutin glycoside presents a significant decrease in anti-radical activity.

Hydroxyl radicals ($\cdot OH$) have a high oxidative activity and severely damage biomolecules susceptible to their action. Hydroxyl radicals are produced in cells during processes such as phagocytosis, during the transformation of PGG_2 to PGH_2 in the biosynthesis of prostaglandins, they are also produced by antineoplastic agents, fungal phytotoxins, ionizing irradiation and the decomposition of lipid hydroxyperoxides, among others.

Hydroxyl radicals are considered primary toxics and a source of secondary toxics, due to their highly reactive properties. The damage caused by ·OH radical's toxic effects is diminished by this radical's trapper such as flavonoids. In 1987, Hussein and collaborators [32], evaluated the trapping activity of ·OH radicals by several flavonoids and found that this activity decreases in the following order: myricetin > quercetin > ramnetin > morine > diosmetin > narinjenin > apigenin > catechin > 5,7-dihydroxy-3',4',5'-trimethoxyflavone > robinin > kaempferol > flavone.

As in the cases mentioned above, the flavonoids studied present different properties of ·OH capture according to their structure. The most active compounds are the flavonols. It has been shown that the presence of hydroxyl groups on the B-ring, especially on the C−3' exerts direct action on the free radical trapping effect. Thus, when the number of hydroxyl groups decreases, the capacity to eliminate the ·OH radicals is rapidly reduced, as in the case of myricetin which traps 50% of the ·OH radicals, while kaempferol traps only 20%. Also important is the presence of the carbonyl function at the C−4 position of the C-ring since, for example, quercetin with carbonyl scavenges 48% of the ·OH radicals, while catechin without carbonyl scavenges 31%.

It was also observed that the hydroxyl radical trapping capacity is independent of the presence of the double bond between C2−C3 of the flavonoid structure. For example flavones such as diosmetin and apigenin had the same ·OH radical trapping capacity as flavanone naringenin. In these compounds the presence of the ·OH hydroxyl group contributed slightly to the trapping of ·OH radicals since the hydroxyl radical trapping capacity in flavones such as apigenin and 5,7-dihydroxy-3',4',5'-trimethoxyflavone was higher than kaempferol.

The antioxidant activity of flavonoids on lipid peroxidation depends from the presence of 3',4'-dihydroxy groups, but its inhibitory action is not increased by the presence of the group C-3−OH, its methylation or conjugation. Hydroxylation is indispensable for the trapping action because, for example, flavones lack a significant trapping effect.

The main source of reactive oxygen species is the mitochondrial energy metabolism chapters 2 and 3. However, the mitochondria themselves have developed chelating mechanisms to inhibit the production of these radicals' chapter 4. Increased electron transport can prevent ROS formation by lowering the O_2 tension in the mitochondrial microenvironment [33].

It has been found that flavonoids are located mainly in chloroplasts. They are considered to suppress lipid photoperoxidation by trapping superoxide and radicals generated during peroxidation. In chloroplasts, hydroxyl radicals ·OH are produced by dismutation of the superoxide anion O^-_2 by means of the Fenton reaction. The trapping of superoxide and hydroxyl radicals in the initiation and termination stages of peroxyl radicals, by flavonoids, constitute their antioxidant action on lipid peroxidation.

The anti-radical properties of flavonoids are mainly directed towards ·OH and $O^{·-}_2$ and towards peroxyl and alkosyl radicals. Furthermore, the antiperoxidative activity of flavonoids is attributed to the chelating capacity of iron due to the high affinity of these compounds for iron ions. The antioxidant capacity of flavonoids against Fe^{2+}-induced linoate peroxidation, due to the ability of flavonoids to donate a hydrogen atom to the peroxyl radical, derived from the autooxidation of fatty acids, was determined by Saija *et al*., 1995 [34] when they evaluated the antioxidant capacity of flavonoids and their interaction with the lipid bilayer.

Furthermore, Silva *et al*. in 2002 [35], by studying flavonoids antioxidant properties, evaluated their ability to reduce ferrylmioglobin, to trap peroxyl radicals generated from the 2,2-diphenyl-1-picrylhydrazyl radical (DPPH·) and 2,2'-azobis(2-amidine-propane) dihydrochloride (AAPH) and to inhibit lipid peroxidation. Lipid liperoxidation was induced by Fe^{2+}-EDTA-ascorbate or AAPH.

Ferrilmyoglobin is a biologically important radical that may be produced after cardiac ischemic reperfusion. The ferrylmioglobin radical (·X-Fe^{IV}=O) is formed in the reaction between metamioglobin (X-Fe^{III}) and H_2O_2, as product of the oxidation of two electrons. In studies, flavonoids were found to effectively scavenge ferryl species. The flavonoids with the highest reducing activity are the flavones quercetin (3,5,7,3',4' − OH), myricetin (3,5,7,3',4',5' − OH), luteolin (5,7,3'4'−OH) and kaempherol (3,5,7,4'−OH). Meanwhile, rutin (3-rut,5,7,3,3'4'−OH), which has the same hydroxyl groups as luteolin but with a rutenoside group at the C-3 position, has low reducing activity. With about half the reducing activity are flavanol catechin and flavanone taxifolin, which have the same hydroxyl groups as quercetin. Considering the above results Silva *et al*. in 2002, concluded that the presence of the double bond 2-3 in the C-ring has an important effect on the reducing activity of flavonoids; furthermore, the inhibition of lipid peroxidation depends on the degree of lipophilicity of the flavonoid molecule that affects its entry into the lipid bilayer and its ability to trap initiation radicals.

It has been proposed that the same structural determinants (Bors' criteria, section 3.1.1.1) that account for the antioxidant action of flavonoids are responsible for the DPPH· radical trapping capacity of flavonoids. A great stability to the resulting radical is given by the *ortho*-catechol group, because when the OH bond is broken, a strong hydrogen bridge is formed between the formed radical and the other OH group, which allows stabilizing the radical and decreasing the enthalpy of dissociation of the O–H bond.

Considering the above, the most effective antioxidant is the flavone myricetin followed by quercetin. However, quercetin has antioxidant capacities like those of catechin and taxifolin; although these flavonoids have the same substitution patterns, taxifolin lacks the (2) determinant and catechin lacks the (2) and (3) determinants. The B-ring is the part of the polyphenol molecule with the highest electron donating capacity [35].

The importance of the presence of the *ortho*-catechol structure on antioxidant activity is supported by the fact that flavonoids such as catechin, quercetin and taxifolin against Fe^{2+}-EDTA-ascorbate-induced peroxidation have a high antioxidant potential. However, kaempferol (3,5,7,4' – OH) also presents high antioxidant activity despite not having the catechol structure. This may be due to the presence of the 2-3 double bonds and the 3 – OH group, where the low oxidative activity of the B-ring may be compensated by the basic structure of the compound.

As evidenced above, the presence of the 2-3 double bond and the 3–OH group is necessary for high antioxidant activity. The presence of the 2-3 double bonds in the C ring gives a high rigidity to the ring and connects the A and C rings in a more coplanar position which supports the difference in antioxidant activity of flavonoid molecules with this structural arrangement. In addition, the 3–OH motif interacts with the B-ring through an intramolecular hydrogen bridge with the 2'- or 6'- hydrogens and this conformational arrangement places the B-ring in the same plane as the A- and C-rings. This planarity of flavonoid molecules is essential for these compounds to effectively enhance the two-electron reduction of ferrylmioglobin ($\cdot X\text{-}Fe^{IV}=O$) to metamioglobin.

The importance of the hydroxyl group of C-3 for a high antioxidant activity was also demostrated by the studies carried out by Arora *et al.* 1998 (6). They observed that the most effective antioxidant in inhibiting metal ion-induced peroxidation was the flavon-3-ol, quercetin. Thus, a slight loss in antioxidant activity occurred when quercetin was replaced by rutin. Considering that metal chelation is fundamental in determining the antioxidant activity of these compounds, the antiperoxidative activity of quercetin and rutin was determined by their ability to chelate Fe ions, thus forming an inert complex incapable of initiating peroxidation.

Myricetin with 6 hydroxyl groups was the most effective flavonoid in inhibiting the oxidation of DHR123 by peroxynitrite with or without bicarbonate. Its level of action was followed by quercetin having 5 hydroxyl groups. However, the flavones rutin, luteolin and kaempferol, with 4 hydroxyl groups and with different hydroxylation patterns, also inhibited the oxidation of DHR123 with or without bicarbonate. This suggests that the antioxidant effect of flavone depends on the number of hydroxyl groups and the presence of a catechol group on the B-ring or a 3−OH.

The reduction of peroxynitrite to nitrite by being oxidized to the corresponding *o*-quinones in a two-electron reaction can be carried out by compounds with catechol groups. Therefore, the presence of an *o*-catechol group (3',4'−OH) in the ring flavonoid structure confers on these compounds the property of being strong electron donors. Antioxidant activity has also been found to be enhanced by myricietin, where a pyragalol (3',4',5'−OH) group is found in the B-ring [36].

Catechins increase the glutathione reductase activity, while decreasing levels of lipid hydroperoxidase (LOOH). The peroxidase activity of lipoxygenase and cyclooxygenase is one of the mechanisms that increase oxidative stress that can be inhibited by flavonoids and phenolic antioxidants *In vitro*. Both enzymes increase oxidative stress as they are part of leucotriene and prostaglandin synthesis, both molecules involved in inflammation. Catechins can also protect against lipid peroxidation produced by oxidative stress initiators such as hydrogen peroxide, t-butylhydroperoxide, 6-hydroxydopamine (6-OHDA), 3-hydroxikinurenine (3-HK), iron (II/III) and ultraviolet radiation [37].

3.2. Redox Chemistry

The antioxidant activity of (-)- epicatechin-3-gallate (ECG), (-)-epicatechin-3-gallate (ECG) and EGCG has been evaluated in several chemically based *in vitro* studies. These compounds' main chemical activity results from either hydrogen atom transfer (HAT) or single electron transfer (SET) reactions, or both, in which hydroxyl radicals are involved. The hydroxyl radicals are part of the B-ring of EC and CGE and the B- and D-ring of ECG and CGE (Fig. **3**).

By the HAT mechanism, catechins act as chain-breaking antioxidants, stopping deleterious reactions such as lipid peroxidation [38]:

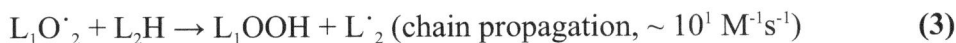

$$L_1H \rightarrow L^{\cdot}_1 \text{ (initiation)} \tag{1}$$

$$L^{\cdot}_1 + O_2 \rightarrow L_1O^{\cdot}_2 \text{ (peroxyl radical formation, } \sim 10^9 \text{ M}^{-1}\text{s}^{-1}) \tag{2}$$

$$L_1O^{\cdot}_2 + L_2H \rightarrow L_1OOH + L^{\cdot}_2 \text{ (chain propagation, } \sim 10^1 \text{ M}^{-1}\text{s}^{-1}) \tag{3}$$

Lipid liperoxidation is a reaction in which hydrogen atoms are abstracted (reaction 1) from unsaturated fatty acids (L$_1$H), and the alkyl radical is produced

($L^{\cdot}_1$) which in turn reacts with molecular oxygen (reaction 2) to give rise to the hydroperoxyl radical ($L_1OO^{\cdot}$). Reaction 2 takes place at constants near to the diffusion limit. The peroxyl radical's abstract hydrogen atoms (reaction 3) from unoxidized lipid substrates (L_2) giving rise to a new lipid alkyl radical ($L^{\cdot}_2$), in this way the chain reaction is propagated. Reaction 3 also produces lipid hydroperoxides (L_1OOH) which are reduced to unstable alkoxyl radicals by transition metal or Fenton type catalyzed reactions and eventually, secondary oxidation products such as aldehydes. Phenolic antioxidants (PhOH) intercept the peroxyl radicals and stop the chain reaction in reaction 3, because this reaction is very slow (ca. $101M^{-1}s^{-1}$):

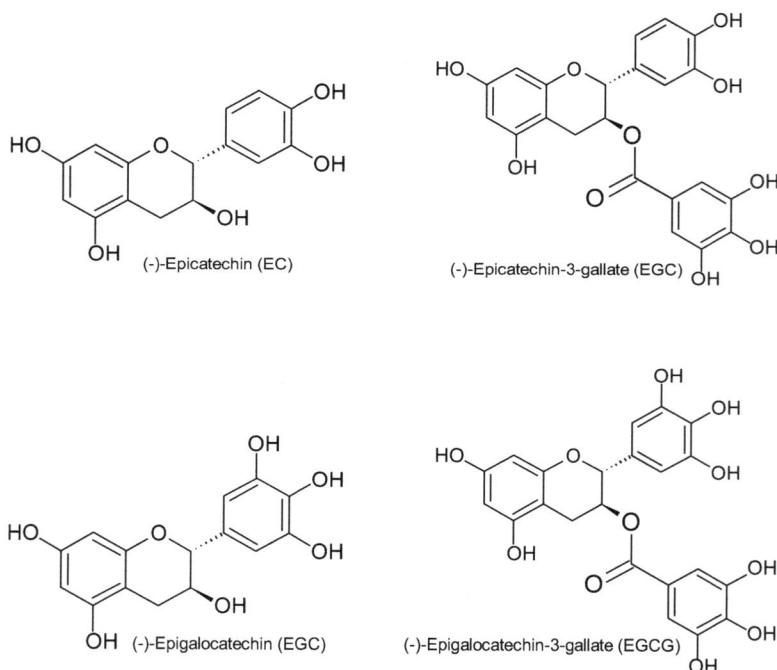

$$L_1O^{\cdot}_2 + PhOH \rightarrow L_1OOH + PhO^{\cdot} \text{ (interruption of chain, } k_4 > k_3) \qquad \textbf{(4)}$$

Fig. (3). Structure of the some polyphenols of tea.

This constant (reaction 4) depends on the bond dissociation enthalpy of the catechin; since the O–H bond of the hydroxyl group is weak, then the reaction with the peroxyl radical is very fast. As the semiquinone radical (PhO$^{\cdot}$) produced is relatively stable, it reacts with unoxidized lipids very slowly. The stability of the semiquinone radical, PhO$^{\cdot}$, is increased by resonance (Fig. 4).

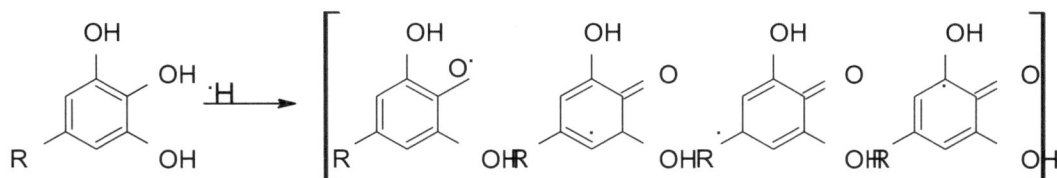

Fig. (4). Resonance stabilization of electrons disappeared in tea catechins.

Catechins are also free radical trappers by the SET mechanism, in which the phenolic radical cations formed subsequently present deprotonation [38]:

$$PhOH + LO_2^{\cdot} \rightarrow PhOH^+ + LO_2 \tag{5}$$

$$PhOH^+ + H_2O \leftrightarrow PhO^{\cdot} + H_3O^+ \tag{6}$$

$$LO_2^- + H_3O^+ \leftrightarrow LOOH + H_2O \tag{7}$$

SET reactions are solvent dependent and their constants depend on the ionization potential (PI) of hydroxyl groups. This involves phenolic antioxidants and peroxyl radicals of lipids originating some products from HAT reactions such as LOOH and PhO$^{\cdot}$. The chain-breaking activity of catechins involves both HAT and SET mechanisms; however, the HAT mechanism predominates and is restricted to the B-ring, also in those catechins with galloyl moieties (ECG, EGCG).

Reactive oxygen species such as superoxide radical, peroxyl radical and chlorine such as hypochlorous acid are trapped by flavanol compounds. Based on the ΔBDE, it has been determined that the order of trapping activity of the major tea catechins is as follows: EC $\leq$ ECG $<$ EGC $<$ EGCG $\leq$ EGCG, utilizing computational methods. This activity order is consistent with the reaction constants obtained in reactions performed with peroxyl radicals in empirical studies.

When the pH is increased, the rate of catechin oxidation also increases. This type of oxidation is auto-oxidative because oxygen reacts with folate anions. It is a thermodynamically favorable but kinetically unfavorable reaction, producing new orbitals containing electrons with the same quantum number, thus not complying with the Pauli exclusion principle. However, essential catalysts in this process are transition metals such as iron and copper which initiate the phenolic oxidation reaction. This reaction also produces O$_2^{\cdot-}$ or under acidic conditions, hydropexyl radical HO$_2^{\cdot}$ (reaction 9), which is also reduced to hydrogen peroxide (reaction 11).

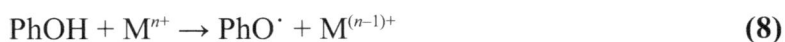

$$PhOH + M^{n+} \rightarrow PhO^{\cdot} + M^{(n-1)+} \tag{8}$$

$$M^{(n-1)+} + O_2 \rightarrow M^{(n-1)+} + O^{\cdot-}_2 \qquad (9)$$

$$PhO^{\cdot} + O_2 \rightarrow QPh + O^{\cdot-}_2 \qquad (10)$$

$$O^{\cdot-}_2 + PhOH \rightarrow PhO^{\cdot} + H_2O_2 \qquad (11)$$

In aqueous solution without the addition of iron or copper, rapid phenolic oxidation also takes place. These metals are known to be contaminants in buffers, culture media, chemical reagents, and solvents. By removing the metal with desferrioxamine, it has been shown that iron catalysis is essential for the oxidation of catechol. Catechol autoxidation was inhibited by catalase, superoxide dismutase (SOD) and the addition of diethylenaminopenta acetic acid under slightly basic pH conditions of 8.0.

Reactive oxygen species are produced in the metal-catalyzed catechins oxidation. These reactions produce highly electrophilic species, the semiquinone and quinone radicals, which react with the free -thiol compounds to generate stable conjugates. The reduction of transition metals ($Fe^{3+} \rightarrow Fe^{2+}$, (reaction 5)), is known to be coupled to the oxidation of catechol and galloyl and the products of this reduction catalyze lipid hydroperoxides and the formation of lipid alkoxyl ($LO^{\cdot}$) and hydroxyl ($^{\cdot}OH$) radicals from the hydrogen peroxide decomposition.

The deleterious effect of metal and phenolic compounds association is related to factors such as changes in pH, among others, but it does not always occur. Under acidic conditions, the ferrous catecholate- and gallate-complexes are less stable, release ferrous ions (Fe^{2+}) and semiquinone radicals that disintegrate. While at neutral pH the metal-phenolic complexes are more stable [38]. Although complex formation results in metal reduction, however transition metal catalytic activity in these complexes is diminished.

When the catalytic ability of the metal bond is retained, the phenolic compound intercepts the radical species formed in the metal complex. Some of the hydroxyl radical trapping activities of catechins can be explained by the reactions occurring at the specific sites where tradical effect ocurs. Experiments on the redox effects of polyphenols have yielded information on antioxidant and pro-oxidant activity mechanisms of these compounds in biological systems (Fig. **5**).

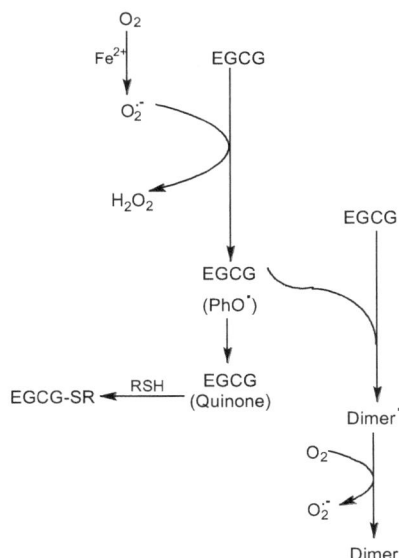

Fig. (5). Oxidation reactions between superoxide, ferric ion and EGCG, a source of oxidative stress. EGCG dimers, EGCG-SR-cysteine conjugates, PhO· radical semiquinone.

4. MECHANISMS OF FLAVONOID ANTIOXIDANT ACTIVITY

Flavonoids' antioxidant action is through mechanisms such as direct trapping of free radicals, enzyme inhibition and chelation of metal ions involved in the production of free radicals and α-tocopherol production.

The direct trapping of free radicals is the main mechanism of free radical activity. Hydrogen atom transfer (HAT), and single electron transfer (SET) are major mechanisms by which flavonoids trap free radicals. In the hydrogen atom transfer (HAT) mechanism, the structural requirements for the antioxidant activity of flavonoids include the *ortho*-dihydroxy substitution in the B-ring, the C2-C3 double bond, and the C-4 carbonyl group in the C-ring. In the HAT mechanism, the hydroxy groups donate hydrogen to the radical, stabilizing the radical and originating a relatively stable flavonoid phenoxyl radical (Fig. **6**). The flavonoid phenoxyl radical produced reacts with a second radical (RO·), thus acquiring a stable quinone structure.

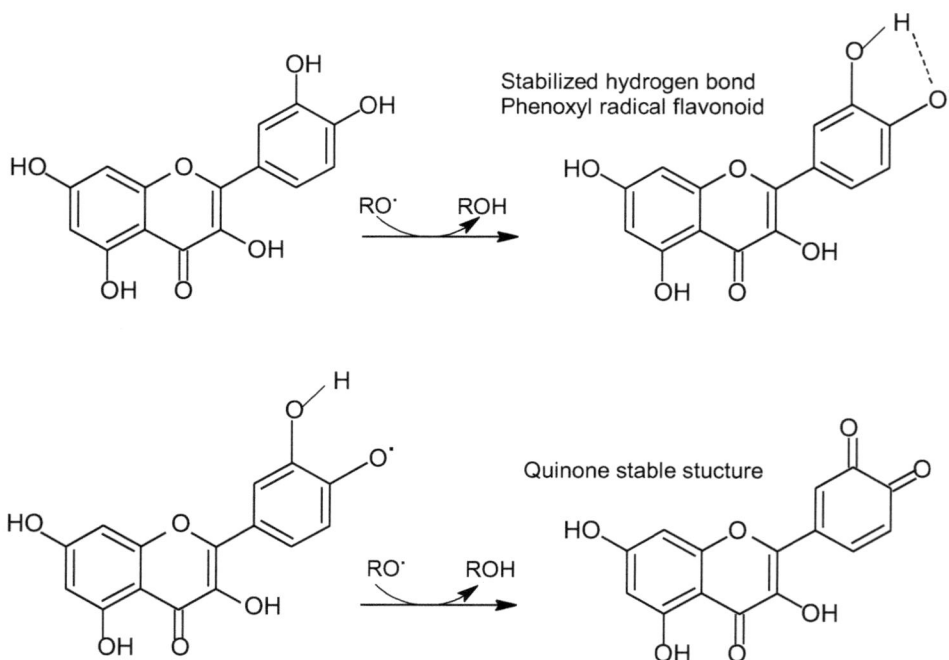

Fig. (6). Antioxidant action mechanism of flavonoids 3', 4'-diOH.

In monohydroxyflavones where hydrogen atom donation-HAT by other hydroxyls is not possible, the electron donating mechanism-SET may be valid. For 3−OH and/or 5−OH hydroxyflavones, the strong hydrogen bonding of its OH group with the oxygen atom of the C-4 carbonyl prevents efficient deprotonation and thus radical scavenging activity by hydrogen atom donation. The mechanism for the antioxidant action of C3−OH or C5−OH hydroxyflavones is illustrated in the following figure:

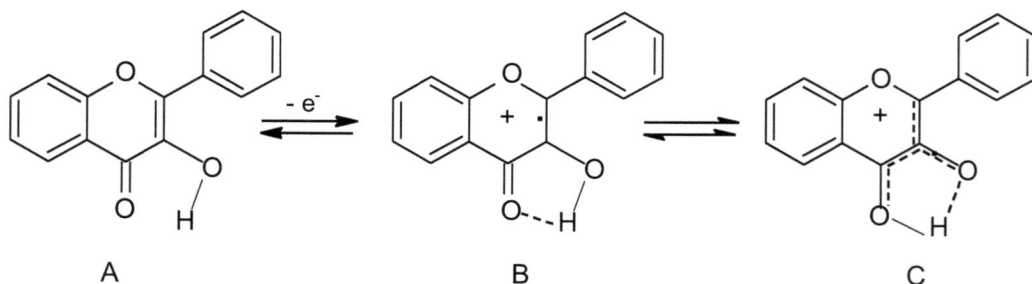

Fig. (7). Mechanisms of antioxidant action of hydroxyflavones C3-OH o C5-OH.

In Fig. (**7**), A corresponds to the structure of 3-hydroxyflavone, which is neutral; the initial cation resulting from electron abstraction from the neutral molecule corresponds to B and the most stable tautomeric form is C, which results from the initial cation B and proton transfer from the C−OH to the C-4 carbonyl group.

During the reduction of hydrogen peroxide (Fenton reaction) [39], reactive oxygen species such as the highly aggressive hydroxyl radicals ·OH are produced (reaction 12) with the contribution of iron(II) and copper(I) which have an important role in oxygen metabolism and the formation of free radicals.

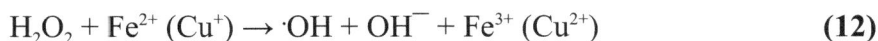

$$H_2O_2 + Fe^{2+} (Cu^+) \rightarrow {\cdot}OH + OH^- + Fe^{3+} (Cu^{2+}) \qquad (12)$$

Several flavonoids effectively chelate trace element ions, such as Fe^{2+} and Cu^+, by binding the metal ions to the flavonoids at the 3',4'-diOH group on the B-ring. The C-3 − OH and C-5 − OH groups in addition to the C-4 carbonyl group of the flavonoid molecule also contribute to the chelation of metal ions (Fig. **8**).

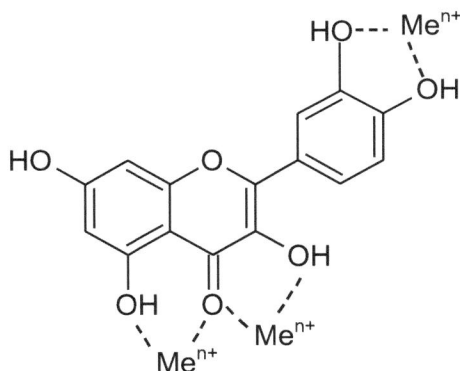

Fig. (8). Sites to connect trace elements.

Furthermore, as free radicals are trapped directly and metal ions are chelated to inhibit their pro-oxicant actions, flavonoids also act as an antioxidant through pro-oxidant enzyme inhibition. This mechanism may be responsible for its *in vivo* effects [28].

4.1. Position and Number of Groups OH

Considering the effect of the number and position of the -OH groups of flavonoid molecules on their antioxidant activity, it has been determined that those flavonoids that present 3',4'-dihydroxy substitution patterns in the B-ring and/or a hydroxyl group in the C-3 position are more effective in trapping radicals. The presence of the C-5 group enhances the trapping activity while the C2-C3 double bond in the C ring is not an essential requirement for high antiradical activity.

Anthocyanins and their aglycones have antioxidant activity that increases with the number of B-ring hydroxyl groups, as demonstrated by structure-activity relationship-SAR studies. This activity decreases, however, when the hydroxyl groups of the B-ring are replaced by methoxyl groups. Anthocyanidins generally exhibit better antioxidant activity than their glycosylated forms and depending on the type of anthocyanidin, their antioxidant power increases or decreases according to the glycosylation pattern.

Structure activity relationships-SAR studies analyzed the O^{-}_2 trapping kinetics and structural features of flavonoids. The entrapment kinetics of the superoxide radical is mainly determined by the B-ring substituents, while the A- and C-ring substituents have little effect on the O^{-}_2 entrapment constants. The fastest superoxide trappers were flavonoid molecules containing an *ortho*-trihydroxy group (pyrogallol) followed by flavonoids with *ortho*-dihydroxy group (catechol). The O^{-}_2 capture kinetics is decreased when the−OH group adjacent to the B-ring is substituted by methoxyl groups. However, it is important to note that in some studies, the existence of the C2-C3 double bond, the existence of the−OH groups at C-3 and C-5 and the keto group at C-4 are not indispensable requirements for superoxide trapping.

The main sites for superoxide radical interaction with the flavonoid molecule are the pyragalol and catechol groups. In this reaction, phenoxyl radicals are produced which, without involving oxygen substituents at C-3 and C-4 and charge delocalization at C-5, may be stabilized by mesomeric equilibrium to an *ortho*-semiquinone structure. Fig. (9) illustrates the mesomeric stability of the flavonoid phenoxyl radical.

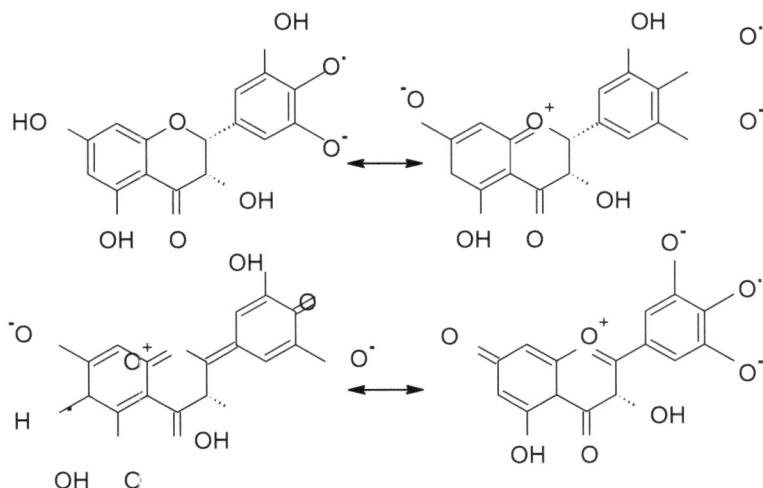

Fig. (9). Mesomeric balance at the phenoxyl radical of flavonoid.

As mentioned above, the increase in the capacity to capture free radicals depends on the number of $-OH$ groups, which in turn is related to the increase in the capacity of abstraction of a hydrogen atom or electron donation. Flavonoid phenoxyl radical stability is achieved by hydrogen bridges, so that the position of the $-OH$ groups, the 3,5-diOH substitution, the catechol moiety in the B-ring and the C-4 keto group are important in achieving stability. Flavonoid phenoxyl radical stability is enhanced by the rearrangement of the $-OH$ groups in the molecule.

The catechol group of the B-ring is the active antioxidant group, as has been demonstrated from SAR studies based on experimental measurements of flavonoid oxidation/reduction potentials. To evaluate the trapping capacity, the average peak oxidation potential ($E_p/2$) of the flavonoids has been utilized.

Both the electrochemical oxidation

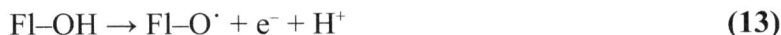

$$Fl–OH \rightarrow Fl–O^{\cdot} + e^- + H^+ \tag{13}$$

as well as the hydrogen atom donating reaction

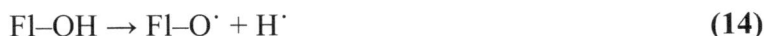

$$Fl–OH \rightarrow Fl–O^{\cdot} + H^{\cdot} \tag{14}$$

implies the same O-H bond breakage.

From the analysis of the oxidation potentials from 23 flavonoids, it was determined that two important characteristics determine antioxidant activity: easy oxidation and lipophilicity. Kinetic measurements and theoretical calculations were employed to estimate the values of n (number of moles of trapped peroxynitrite radicals). The formation of the *ortho*-quinone product (Fig. **10**) was evidenced by MP3 calculations where the value of n was 4 while the experimental value was 3.54.

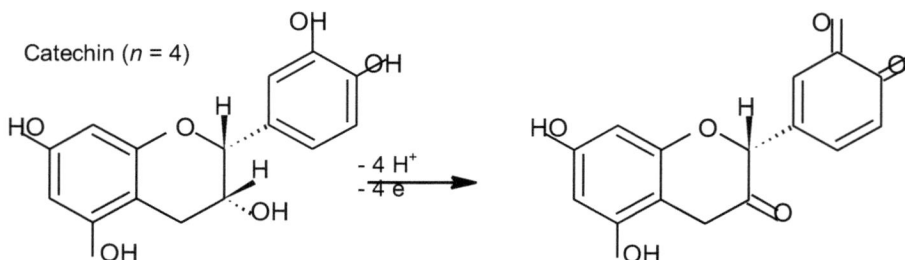

Fig. (10). Catechin ($n = 4$) and *ortho*-quinone product oxidized.

By means of the Ferric Reductive Antioxidant Power (FRAP) method, the antioxidant activity of several flavonoids was also evaluated [28]. By cyclic voltammetry, the oxidation potential of the flavonoids was determined. A correlation between FRAP values and anodic oxidation potential ($r = -0.907$) was established. The most important factors in the determination of antioxidant capacity were the hydroxyl groups, the C2-C3 double bond and the catechol group, 3−OH. The correlation of n OH with oxidation potential ($r = -0.960$) was higher than with FRAP values ($r = 0.908$).

Flavonoids may also contribute to protection against oxidative damage by scavenging the singlet oxygen $^1O_2$, as demonstrated by a kinetic study of the scavenging reaction of ($^1O_2$) by eight flavonoids. For the reaction of $^1O_2$ with flavonoids, the (k_Q) constants were found to increase with increasing substituted OH in the flavone base skeleton.

For the extinction reaction of $^1O_2$ by flavonoids, the catechol or pyrogallol structure in the B-ring is essential. Flavonoids with low E_{pa} values have high reactivity. For flavonoids with C2-C3 double bonds, log k_Q correlates well with E_{pa} ($r = -0.99$). However, for flavonoids without C2-C3 double bonds, a deviation from the linear correlation was observed.

4.2. Formation Heat of Flavonoid Radicals (ΔH_f)

Radical ΔH_f is the difference between the heat of formation of flavonoid parent and its respective radical, which is obtained by hydrogen atom abstraction from the indicated−OH group. This ΔH_f represents the stability of a phenoxyl radical with respect to its parent flavonoid and is employed to compare the different positions of the groups in an individual flavonoid, as well as between flavonoids which allows explaining some of flavonoids' antioxidant activities experimentally. Independently of flavonoid subclass or substitution pattern, the ΔH_f calculation for the reaction FlOH → FlO˙ + H˙ presents better results for a high activity molecule. At a low ΔH_f value, phenoxyl radical is more stable and thus more active antioxidant.

Considering the above, the effect of flavonoids on the inhibition of low-density lipoprotein (LDL) oxidation *in vitro* has been investigated. A linear correlation was found between the experimentally obtained antioxidant activity values and the calculated ΔH_f values. The calculated ΔH_f indicates that a better result was obtained in the donation of hydrogen atoms from the 3−OH group to the C-3 group, followed by an−OH from the B-ring.

Utilizing the ΔH_f and spin densities, the flavonoid structure-antioxidant activity relationship has also been studied. It has been found that for lower ΔH_f values, flavonoids with hydroxyl groups at C-4' and/or C-3' are more active and that according to the spin density map analysis, the unsaturation between C2-C3 enables resonance stabilization of the formed radicals.

In the hydrogen transfer from the flavonoid to the thyroxyl radical, the heat of reaction was ($\Delta\Delta H_f = (\Delta H_{f(flavonoid)} - \Delta H_{f(tirosin)})$), as in the case of the reaction of the thyroxyl radical restored by flavonoids:

$$TyO˙ + FlOH → TyOH + FlO˙ \qquad (15)$$

When studying the influence of structure-related parameters of several flavonoids on lipid peroxidation, such as the hydration energy E_{HYDR}, ΔH_f, and the energy of the least occupied molecular orbitals (E_{LUMO}), it was found that flavonoids with the highest antioxidant capacity possess the highest number of−OH groups and have the highest E_{HYDR} values, while flavonoids with low antioxidant activity have the lowest E_{HYDR} values. Interestingly, the E_{HYDR} parameter indicates flavonoids' hydrophilic properties.

4.3. Bond Dissociation Energy (BDE) And Ionization Potential (IP)

Density Functional Theory (DFT) studies to determine the antioxidant capacity of various flavonoids established that the calculation of BDE is important for hydrogen atom transfer-HAT studies and the calculation of PI is important for single electron transfer-SET studies. Hence, a high ability to donate a hydrogen atom from the hydroxyl group and thus trap free radicals occur with low BDE values. And a decrease in the electron transfer rate between the antioxidant and oxygen, which would lead to a reduction in the prooxidant activity of the antioxidant, occurs when there is a relatively high PI value. A relatively low BDE value for the O–H, has the catechol group in the B-ring.

Free radical scavenging, anion-derived flavonoids are more active than neutral molecules. Descriptors such as bond energy, dissociation constants, absolute hardness and partition coefficient can be utilized to establish the relationship between the acidity of the–OH group and the biological activity of flavonoids.

The second-order reaction constants (k_2) for galvinoxyl radical reduction by flavonoids depend mainly of OH group configuration in the B- and C-ring of the flavonoid (according to the BDE values for the O–H bond). This was determined in stoichiometric and kinetic studies of the hydrogen-donating capacity of various flavonoids using electron spin resonance (ESR). When flavonoids are oxidized to *ortho*-quinones or extended to *para*-quinones, high reaction constants and high stoichiometric reactions are obtained. For example, between $\log(k_2)$ and stoichiometric reaction, a moderately high correlation was found ($r = 0.818$).

Utilizing the MP3 method, the phenolic O–H and BDEs were determined for the catechins C–H. The hydrogen at the C-2 position can be abstracted by the free radical since the BDEs were lower for the C–H of the catechins at the C-2 position than for the BDEs of the C–H at the phenolic sites.

The effect of pH on the antioxidant properties of 22 hydroxyflavones (TEAC assay) was studied and it was found that pH influences the antioxidant capacity related to the deprotonation of the hydroxyl group, leading to a higher oxidation potential for the formation of the deprotonated forms. With the experimental calculations of BDE for the O–H and IP of the protonated and deprotonated forms of hydroxyflavones, it was determined that the BDE parameter mainly indicates the easy electron donation and the IP is influenced by deprotonation.

The above indicate that after deprotonation, the radical trapping capacity increases because electron donation is easier. Based on the BDE and IP values of several flavonoids, it was determined that epicatechin, luteolin and taxifolin act as hydrogen donors. While apigenin, luteolin, kaemferol act through the simple electron transfer mechanism, because their low IP values are determined by the planar conformation and the electronic delocalization of the extended π-bond between adjacent rings.

Trouillas *et al.*, 2006 [40], determined the specificity of the 3–OH group in quercetin and taxifolin antioxidant action. Only when the C2-C3 double bond is present, the 3',4'-diOH and 3 − OH groups exert a significant effect on the antioxidant action of these flavonoids, as demonstrated from BDE values analysis for the OH sites in these molecules. The C-ring opening observed in flavonol degradation during metabolism is due to the high spin density that the 3−OH radical of quercetin has on the C-2 atom, as indicated by the spin density distribution in the radicals formed when H removal from each OH site occurs, in both flavonoid molecules.

In a conjugated system, if the unpaired electrons are highly delocalized, the energy of a free radical can be effectively lowered. In this case, the spin density is a suitable descriptor to characterize the free radical stability. Higher antioxidant activity occurs in derivatives with *ortho*-hydroxy-amino groups than in those with monohydroxy or *ortho*-dihydroxy group.

As mentioned in section 4 of this chapter, the free radical scavenging process by flavonoids is mainly through the HAT and SET mechanisms.

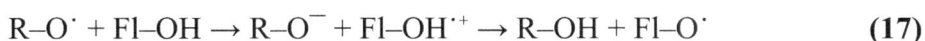

$$R–O^{\cdot} + Fl–OH \rightarrow R–OH + Fl–O^{\cdot} \tag{16}$$

$$R–O^{\cdot} + Fl–OH \rightarrow R–O^{-} + Fl–OH^{\cdot +} \rightarrow R–OH + Fl–O^{\cdot} \tag{17}$$

In reaction 16, the mechanism of hydrogen atom transfer-HAT may be characterized by BDE of the OH groups and in reaction 17, the electron transfer-SET may be measured by the PI. As the ability of the flavonoid to trap the radical is higher, the lower these parameters will be. As mentioned above, the main descriptors for the antioxidant capacity of flavonoids are ΔH_f, BDE of the OH groups, and IP.

Quercetin and rutin can inhibit free radicals in cells in three situations: in the initiation by interaction with superoxide ion, in the formation of hydroxyl radicals, by iron ion chelation and in the peroxidation of lipids by reaction with peroxyl radicals [41].

4.3.1. Quercetin

The reaction of the OH group located on the C-ring is mainly responsible for quercetin reactivity. However, the 4'−OH is the appropriate site for the action of the radical, as demonstrated by spin density, BDE, and density functional theory-DTF studies. Furthermore, in the oxygenolysis reaction, the copper atom has an oxidative character towards quercetin.

Values of geometric parameters have been determined such as (bond length and bond angle) results of the optimization of the geometry of neutral quercetin (fig. 11).

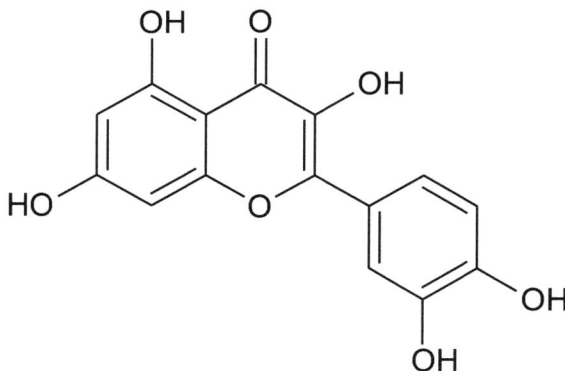

Fig. (11). Quercetin structure.

Dhaouadi *et al.*, 2009 [42], has conducted studies on the interaction of quercetin with two ROS especially the attack of the radicals O^-_2 and $\cdot OH$ to the 4'$-$OH position of the B ring.

Following the binding of the superoxide radical O^-_2 to the 4'$-$OH position, a wide separation between $H_{4'}$ and $O_{4'}$ is observed. This distance increases by 0.55 Å with respect to the $O_{4'}-H_{4'}$ bond in unaltered quercetin. Also, the hydrogen $H_{4'}$ binds back to the oxygen atom O_A of the radical (1.044Å) by a weak bond. Notably, this atom is also bonded to oxygen $O_{4'}$ by a strong hydrogen bond (1.515 Å).

The binding of the $H_{4'}$ atom to the radical produces a perturbation that propagates throughout the molecule. A variation of the valence angles for $C_3-C_4-C_{5'}$, $O_{3'}-C_{3'}-C_{4'}$, $C_{3'}-O_{3'}-H_{3'}$ and $C_{4'}-O_{4'}-H_{4'}$ of 3.2°, 5.1°, 5.6°, and 14.1° respectively, has been registered in the B ring where this perturbation is more intense. Analyzing the reduction of the valence angles $O_{3'}-C_{3'}-C_{4'}$ and $C_{3'}-O_{3'}-H_{3'}$ together with the shortening of the $C_{4'}-O_{4'}$ bond, it is observed that these mainly contribute to strengthen the hydrogen bridge $O_{4'}\cdots H_{3'}-O_{3'}$ (1.975Å), because of the lengthening of the C_4-O_4 bond (0.02Å) and the opening of the valence angle $O_5-O_5-H_4$ (2.9°).

The binding of the $\cdot OH$ radical at the 4'$-$OH position of the molecule induces a transfer of the $H_{4'}$ atom to the O_A atom. Similarly, a lengthening of the distance between the $O_{4'}$ and $H_{4'}$ atoms of 0.93 Å was observed. Also, $H_{4'}$ was bonded to the oxygen atom O_A by a strong covalent bond, which was demonstrated because the $O_A-H_{4'}$ distance was 0.980 Å. This demonstrates that the ·OH hydroxyl radical is a stronger oxidant than the O^-_2 superoxide radical.

The high reactivity of hydroxyl radical ·OH with respect to the O^-_2 superoxide anion radical was demonstrated because the $C_{4'}-O_4$ bond distance was reduced and its value approaches the carbon-oxygen double bond distance. This distance shortening demonstrates a transfer of one hydrogen from $O_{4'}$ to O_A. Similarly, the bond strength between $H_{3'}$ and $O_{4'}$ was reduced, in relation to the results observed after superoxide radical binding. The B-ring was affected more than the C-ring by the perturbation originated from hydrogen abstraction. The length of the $C_{4'}-C_{5'}$ bond presented a variation by 0.05Å. This perturbation also affected the strength of the $O_4...H_5$ hydrogen bond. The above results demonstrate that the attack of the ·OH hydroxyl radical weakens the hydrogen bond, while this bond is strongly strengthened by the O^-_2 superoxide radical binding.

The phenolic sites (3−OH, 5−OH, 7−OH, 3'−OH, and 4'−OH) presented the following bond dissociation energies (BDEs): 76.2, 89.9, 81.5, 78.9 and 67.6 Kcal/mol [42], corresponding to the following order 4'−OH < 3'−OH < 3'−OH < 7−OH < 5−OH. The phenolic hydrogen at the 4'−OH site can be more easily detached than at the 3−OH site, because the BDE value at the 4'−OH site was lower than that at the 3−OH site by 8.6 kcal/mol.

These results are like those obtained in other studies, being the catechol group the most important moiety in phenolic acids and flavonoids. The catechol group is more reactive with free radicals at the 4−OH site. When comparing the BDE values of the hydrogen atom abstraction at the 4−OH and 3−OH positions of 3,4-dihydroxybenzoic acid (3,4-DHBA) with the BDE of the 4'−OH and 3'−OH positions of quercetin, the order 3'−OH < 3−OH and 4'−OH < 4−OH is obtained. This demonstrates that the antioxidant activity of quercetin is higher than that of 3,4-DHBA. By comparison of the BDEs obtained for the 4'−OH and 3'−OH positions of the catechol group of quercetin, 3,4-dihydroxypyruvic acid (3,4-DHPPA) and 3,4-dihydroxycinnamic acid (3,4-DHCA), the antioxidant activity presented the following order: quercetin > 3,4-DHPPA > 3,4-DHCA > 3,4-DHBA. Demonstrating that the ability to remove an electron from the A and C rings in quercetin (Fig. **11**) is greater than that of the side chains of 3,4-DHPPA and 3,4-DHCA. This is due possibly to the $C_2 = C_3$ unsaturation in the C-ring and the $O_3-H_3...O_4$ hydrogen bond, the presence of delocalized π electrons in the A-ring, and the presence of another hydrogen bond between O_5H_5 and O_4.

In the following reactions:

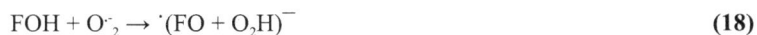

$$FOH + O^-_2 \rightarrow ·(FO + O_2H)^-$$ (18)

$$FOH + \cdot OH \rightarrow \cdot (FO + H_2O) \tag{19}$$

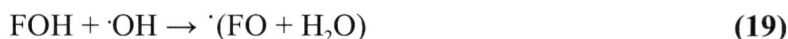

Quercetin is represented by FO. At the 4'−OH position the attack of the radical is present. The associated ΔH_f enthalpies for each reaction correspond to the total enthalpy difference between the products and reactants in each case. For reaction 18, the ΔH_f is about -52 kcal/mol, while for reaction 19 it corresponds to - 47 kcal/mol. Since these reactions are exothermic, they promote these products formation.

In reaction 18 and in reaction 19 the variation of the $O_{4'}-H_{4'}$ bond length affects the surface energy potential (SEP) of the interaction between quercetin and the $O^{\cdot-}_2$ radical anion or between quercetin and the $\cdot OH$ radical. On the other hand, $C_{4'}-O_{4'}-H_{4'}$ angle variation only affects reaction 18. This variation affects the strength of the hydrogen bridge $H_{3'}...O_{4'}$, and electron delocalization. A variation of 14.1Å for $C_{4'}-O_{4'}-H_{4'}$ is registered for reaction 18 and for reaction 19 the $H_{3'}...O_{4'}$ initial distance 1.975 Å becomes 2.047Å.

Analysis based on NBO (natural bonding orbital) charges and spin density studies were utilized to identify the fragments of the following reactions:

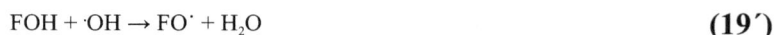

$$FOH + O^{\cdot-}_2 \rightarrow FO^- + \cdot O_2H \tag{18'}$$

$$FOH + \cdot OH \rightarrow FO^\cdot + H_2O \tag{19'}$$

Comparing the charges on the products associated with the 18' reaction, the FO fragments are negatively charged, whereas $\cdot O_2H$ is a radical. Since this reaction has no transition states, the transference can be explained by two mechanisms: In reaction 20 quercetin binds to the superoxide anion radical which, by the transfer of a hydrogen atom, is converted to HO_2^- followed by electron transfer to produce the $FO^\cdot$ fragment.

$$FOH + O^{\cdot-}_2 \xrightarrow{\cdot H} FO^\cdot + HO_2^- \xrightarrow{e^-} FO^- + \cdot O_2H \tag{20}$$

In reaction 21, proton transfer from quercetin to the superoxide anion radical occurs.

$$FOH + O^{\cdot-}_2 \xrightarrow{H^+} FO^- + \cdot O_2H \tag{21}$$

The NBO analysis of reaction 19' showed that the hydrogen transfer has a positive charge and is not associated with the spin, which indicates that it is a proton transfer. During the transition state the spin densities of the FO· fragment increase and those of the hydroxyl radical decrease, this means that the proton and electron are transferred in the same direction from the quercetin to the hydroxyl radical, because there is a coupled transfer of protons and electrons. The 19' reaction has a reaction constant of $1.5 \times 10^{11} s^{-1}$.

4.3.2. Rutin

Rutin is a type of flavonoid glycoside (Fig. **12**) known as vitamin P, which due to its properties is known as a multipotent antioxidant. Rutin is a type of flavonoid glycoside (Fig. **12**) known as vitamin P, which due to its properties is known as a multipotent antioxidant. Some research has demonstrated rutin's antioxidant activity as a superoxide and hydroxyl radical trapping agent.

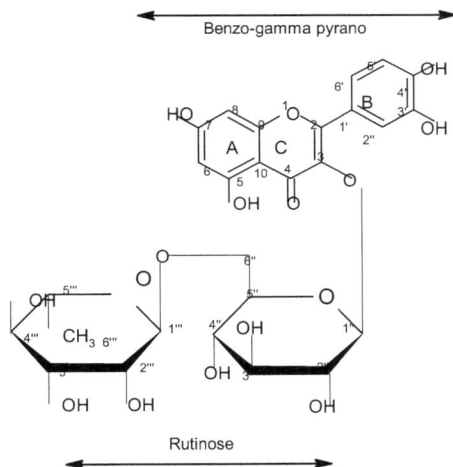

Fig. (12). Chemical structure of rutin.

The RO· radical formed when reacting rutin with the hydroxyl radical is stable and the reaction thermodynamically favorable. Radicals formed by abstraction of a hydrogen atom in rutin are (3'−OH, 4'−OH, 5−OH, 7−OH) whose BDE values are: 72.4; 74.1; 95.6 and 84.2 respectively, which have the following OH group order: 3'−OH < 4'−OH < 7−OH < 5−OH [43]. Thus, the highest BDE value, 95.6 kcal/mol, corresponds to the 5−OH group, since the internal hydrogen bonding is thermodynamically very unfavorable. The transfer of H from the 3' position of the B-ring is energetically more favorable. Hence, the B-ring of rutin is the most important site for antioxidant capacity.

When the spin densities of the radicals formed from the B and A rings were compared, it was determined that after abstraction of the hydrogen atom, the stabilization of the radical was due to the delocalization of the π electron. The relationship between radical reactivity and electron delocalization was determined by high occupancy molecular orbital (HOMO). Consequently for the four ·OH radicals formed, the HOMO form is similar and does not present variations in antioxidant activity among them.

The spin population on the O after abstraction of the H atom and on the neighboring C atoms is slightly more delocalized for the radicals from the B ring (3'−OH and 4'−OH) than for those from the A ring. Thus, the spin density on the O atom is 0.24 and 0.33 for the 3'−OH and 4'−OH sites respectively and for the 7-OH and 5-OH radicals is 0.62 and 0.71.

Radical formation is facilitated by the delocalization of the spin density of the radical and thus the BDE is decreased, which implies that for rutin the BDE is lower in the B-ring than in the C-ring and that the site for the binding of the ·OH radical is the 3'−OH. However, the attack of the ·OH radical is directly on the oxygen atom adjacent to C-3', with the production of an intermediate-1. On the other hand, when the superoxide radical reacts with the rutin at the 3'−OH and 4'−OH sites, the intermediates-2 and -3 respectively are formed.

The homolytic cleavage of the 3'−OH bond yields the intermediate-2 forming H_2O_2 and the intermediate-3 is obtained by heterolytic cleavage of the 4'−OH bond yielding the anionic radical of rutin. Thus, rutin is oxidized by the radicals giving rise to a more stable and less reactive radical. Rutin's radical-trapping capacity may be due to its xanthine oxidase enzyme inhibitory activity.

4.4. Chelation of Metallic Ions

Other mechanisms exist to determine the flavonoids antioxidant capacity, such as flavonoid's ability to chelate metal ions, resulting in their inactivation to participate in free radical generation. The flavonoids interactions with iron and copper ions have been studied. Stoichiometric ranges of the metal-flavonoid complex: 1:1, 1:2, 2:2, 2:3 was observed. Stoichiometry 1:2 is the most utilized. The metal-binding sites on flavones are mainly the 5-hydroxy and 4-oxo groups. Furthermore, the *ortho*-catechol group can be chelated. In flavonoids the structure-activity relationship (SAR) involves the reduction of iron (III) and copper (II) ions. Flavonoid antioxidant activity on superoxide radicals initiates with redox flavonoid-Fe^{3+} complex formation.

The catechin and taxifolin acidity constants and the copper (II) complex formation constants were determined by spectrophotometry and/or potentiometry. Also, were determined the values of the partition coefficient (log P). The anionic forms of flavonoids are predominant in metal chelation. A removal of a proton in its molecule is important for flavonoid antioxidant activity since chelation occurs at least through a non-protonated ligand.

5. INHIBITION OF PROOXIDANT ENZYMES MECHANISMS

Reactive oxygen species (ROS) production is directly influenced by the action of prooxidant enzymes such as xanthine oxidase, lipoxygenase, mitochondrial succinoxidase, cyclooxygenase, NADPH oxidase and microsomal mono oxygenase.

Flavonoids can inhibit redox enzymes such as (monooxygenases, cyclooxygenases, lipoxygenases, xanthine oxidase, and NADPH oxidase) due to the reaction that occurs within the enzyme's active site, between some flavonoids and free radicals generated there.

Catechins also induce phase II enzymes for detoxification or antioxidant enzymes such as superoxide dismutase (SOD), catalase and glutathione peroxidase (GPx).

5.1. Xanthine Oxidase

Oxidation reaction of hypoxanthine and xanthine to uric acid and superoxide radical is catalyzed by xanthine oxidase (XO) enzyme. Studies on xanthine oxidase inhibition by flavonoids reveal that an important requirement for flavonoid activity is the C-2 and C-3 double bond as the planar center of the flavone, as well as the presence of hydroxyl groups at C-5 and C-7. Thus, flavones with 7−OH group and catechol or pyragalol group in the B-ring inhibit XO more efficiently (Fig. **13A**).

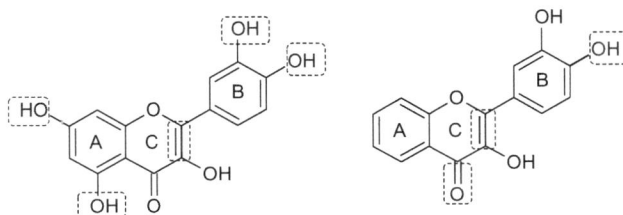

Fig. (13). Structural requirements for: A) Xanthine oxidase inhibition; B) Lipooxygenase inhibition.

Numerous studies on flavonoid SARs as XO inhibitors and superoxide radical scavengers found that the flavonoid planar structure is relevant for the inhibition of XO. The coplanarity of ring B with rings A and C is due to the conjugation of the C2-C3 double bond. XO inhibitory activity by flavonoids is also favored by the presence of the hydroxyl groups at C-5 and C-7, but is diminished by the hydroxyl group at C-3. However, the presence of hydroxyls at C-3' and C-3 were essential for superoxide radical trapping activity. By means of molecular mechanism (MM+) and AM1 methods, it was possible to study the effect of steric, electrical, and hydrophobic properties of flavonoids on XO inhibition. Flavonoid molecule geometry has been shown to be essential for its activity as an XO inhibitor. Thus, flavonoid cannot inhibit the XO enzyme when the torsion angles between the C and B rings are greater than 27°. The inhibitory activity is enhanced both by the small size of the flavonoid and its high hydrophobicity (log *P* values).

Xanthine oxidase (XO) is another enzyme inhibited by catechins. Xanthine oxidase contains two active sites each with one molybdenum atom, as well as two distinct Fe/S I and II centers, and one FAD molecule [44, 45]. During the oxidation of xanthine oxidase reduced by molecular oxygen, there are two distinct kinetic phases in which the respiratory rate in the fast phase is 10 times higher than in the slow phase. In the case of the enzyme reduced by 6 electrons, the fast oxidation phase reflects five electron elimination and a slow phase where the remaining electrons are eliminated. When xanthine oxidase is completely reduced, all enzyme molecules are rapidly converted to a state of electron reduction which is slowly re-oxidized due to oxygen reaction with a small amount of present flavin radical.

In experiments realized by Olson *et al.*, in 1974 [45], to examine xanthine oxidase reduced reaction with molecular oxygen; they found that flavine is the oxygen attack site even for reduced enzyme molecules in an electron. At low pH, considerable amounts of semiquinone flavine are generated during the rapid phase of the oxidation reaction and decrease considerably at high pH. However, at high pH values, electrons initially found in Fe centers are removed by oxygen reacting with a small amount of semiquinone. Flavin semiquinone reaction with oxygen occurs in the slow phase of the oxidation reaction, suggesting that the intramolecular transfer rate from the iron and molybdenum centers to the flavin is faster than the oxygen reaction rate.

To explain the oxidation, the simplest model assumes that the $FADH_2.O_2$ complex is broken to produce a mixture of O^-_2 and H_2O_2 [44] and the proportion is kinetically determined by the different transfer rates of 1- 2 electrons from flavin completely reduced to oxygen. In the case of the one-electron reduced enzyme molecules, $FADH_2$ was not formed; therefore, a low respiratory rate was observed.

Although in the fast oxidation phase the yield of one-electron reduced enzymes indicates that one-electron transfer and large amounts of $O^{.-}_2$ are indispensable; however, during this reaction reduced enzyme molecules are produced but without the production of $O^{.-}_2$, perhaps since oxygen rapidly binds to the totally reduced flavin and forms the $FADH_2.O_2$ complex (reaction 22). Thus, an electron is quickly transferred to the oxygen.

An intramolecular electron transfer from the reduced iron and molybdenum centers generates fully reduced flavin when the enzyme has more than two electrons. A second electron can be transferred from $FADH_2$ to $O^{.-}_2$ and thus form H_2O_2 (reaction 25) if this process is faster than the rate of diffusion of $O^{.-}_2$ out of the active center.

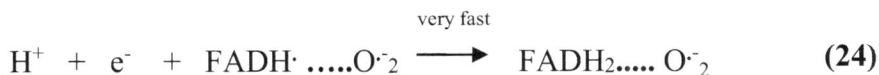

$$FADH_2 \;+\; O_2 \;\rightleftharpoons\; FADH_2{\cdot}O_2 \qquad\qquad (22)$$

$$FADH_2{\cdot}O_2 \;\longrightarrow\; FADH{\cdot} \,.....O^{.-}_2 \;+\; H^+ \qquad\qquad (23)$$

$$H^+ \;+\; e^- \;+\; FADH{\cdot} \,.....O^{.-}_2 \;\xrightarrow{\text{very fast}}\; FADH_2 .....\,O^{.-}_2 \qquad (24)$$

$$FADH_2 .....\,O^{.-}_2 \;\xrightarrow{\text{very fast}}\; FADH{\cdot} \;+\; H_2O_2 \qquad\qquad (25)$$

$$FADH{\cdot} \,.....\,O^{.-}_2 \;\xrightarrow{\text{fast}}\; FADH{\cdot} \;+\; O^{.-}_2 \qquad\qquad (26)$$

Until two electron-reduced enzyme molecules are generated, this process continues. However, after the transfer of one electron to oxygen, no $FADH_2$ is generated because the remaining electron cannot react with either $O^{.-}_2$ or O_2 since it is located in the reduced iron center. Nevertheless, $O^{.-}_2$ is produced rapidly only in the case of two-electron reduced enzymes (reaction 26) [45].

5.2. Lipoxygenase

The enzymatic peroxidation of lipids is catalyzed by prooxidant enzymes such as lipoxygenases (LOX), they can also initiate non-enzymatic lipid peroxidation and participate in other oxidative processes. In SAR studies, evaluating the inhibitory activity of flavonoids on 15 lipoxygenases (15-LOXs), it was found that structural features in the flavonoid molecule such as: 1) the catechol group in the B or A rings, 2) the 4-carbonyl group in the C ring (Fig. **7B**) and 3) the presence of delocalized electrons involving the C2-C3 double bond, considerably enhance the inhibitory action. On the contrary, the presence of 3−OH in the C-ring decreases the inhibition of 15-LOX. Significantly, the excess of OH groups decrease flavonoid hydrophobicity and affects its positioning in the hydrophobic cavity of the active site of the enzyme, whereas the presence of the catechol group is favorable for flavonoid inhibitory action.

The presence of the catechol group on the B-ring of the flavonoid molecule contributes significantly to the inhibition of 15-LOX induced by LDL oxidation. This oxidation is carried out by quercetin, quercetin 3-glucoside and quercetin 7-glycoside. However, quercetin 4'-glucoside has a greater inhibitory effect. Quercetin glucosides and its aglycone induce LDL oxidation more efficiently than vitamins C and E.

Quercetin and its two glucuronide metabolites (3'- and 4'-) are also inhibitors of xanthine oxidase and trap superoxide radicals such as quercetin-4'-glucuronide performs. The concentration of 0.25 uM corresponds to the k_i for XO inhibition by quercetin-4'-glucuronide. Quercetin glucuronides are moderate inhibitors of soybean lipoxygenase, except quercetin-3-glucuronide in Table 1 [46].

Table 1. Enzymatic kinetic of quercetin inhibitory effects and metabolites on Xanthine Oxidase and Lipoxygenase (Competitive and non-competitive inhibition).

-	Inhibition (K_i µM)	
	Xanthine Oxidase	**Lipoxygenase**
Quercetin	0.2	2.8
3′-methylquercetin-	0.25	4.8
Quercetin-3-sulphate	78	16
Quercetin-3-glucuronide	160	60
Quercetin-7-glucuronide	100	60
Quercetin-4′-glucuronide	0.25	8.4
Quercetin-3′-glucuronide	1.4	6.5

6. MEASUREMENT METHODS FOR FLAVONOID ANTIOXIDANT ACTIVITY

Analytical methods for measuring the antioxidant capacity (AOC) of plant compounds have increased considerably in diversity [47]. There are two main mechanisms by which the antioxidant activity of a chemical compound can be realized: Hydrogen atom transfer - HAT and single electron transfer - SET. According to this, there are different methods, each of which has advantages and disadvantages.

By the Hydrogen Atom Transfer - HAT method, it is possible to measure the capacity of an antioxidant to eliminate free radicals by hydrogen donation. Among these methods are: Oxygen Free Radical Absorbance Capacity - ORAC and Radical Trapping Capacity - TRAP. The ORAC method measures the antioxidant capacity, or the inhibition of oxidations induced for example by peroxyl radicals, in function of time. ORAC values are presented as Trolox equivalents. This method employs as oxidizable protein or substrate beta-phycoerythrin (PE) and as peroxyl radical generator 2,2'-azobis(--amidinopropane) dihydrochloride (AAPH); it can also employ as hydroxyl radical generator a $Cu^{2+} - H_2O_2$ system.

Cao *et al.*, 1997 [48], evaluating the antioxidant and pro-oxidant capacity of some flavonoids through oxygen radical absorbance capacity -ORAC test, determined that this technique has advantages over other methods for the following reasons: The ORAC system takes into account the Low Area Curve (AUC) which allows combining the inhibition time and the degree of inhibition of free radical action, free radical generators or oxidants can be utilized since the antioxidant activity depends on the free radical and the oxidant. In other methods the time or the degree of inhibition may remain fixed.

Flavonoids can act as antioxidants or prooxidants, depending on the source of the free radical and its concentration. For example, they may have antioxidant activity against free radicals, however, in presence of a transition metal act as prooxidants. Both antioxidant activities including $ORAC_{ROO.}$ and $ORAC_{.OH}$, as well as prooxidant activities depend on flavonoid structure [48].

The TRAP method is a test utilized to measure antioxidant capacity- OAC in serum or plasma because it measures non-enzymatic antioxidants such as glutathione, α-tocopherol, ascorbic acid and β-carotene. It is also employed to measure flavonoid interference with peroxyl radical reactions produced by ABAP [2,2'-azobis(2-amidinopropane) dihydrochloride] and a target assay.

Total Oxygen Scavenging Capacity - TOSC method quantifies antioxidant absorption capacity towards hydroxyl radicals, peroxyl radicals and peroxynitrite. The substrate that is oxidized is α-keto-γ-methylbutyric acid (KMBA), which in turn forms ethylene. Thus, the antioxidant capacity is quantified by the ability to inhibit ethylene formation with respect to the control assay. This method also utilizes the area under the curve by obtaining the experimental data during a reaction time, obtaining linear dose-response curves from the reaction kinetics [49].

To evaluate the antioxidant capacity of plant samples, the β-carotene linoleic acid decolorization test is employed. This test is utilized because carotenoids are discolored by autooxidation, peroxyl radical-induced oxidation, light, or heat. Discoloration degree can be decreased by the application of antioxidants that trap free radicals through hydrogen atom donation. In this method, the antioxidant capacity is verified when no volatile organic compounds are produced and instead diene peroxide conjugates are obtained when linoleic acid is oxidized, which originates β-carotene discoloration. Chloroform (1ml) with β-carotene (0.5mg) is added to 25 µl of linoleic acid and 200 mg of Tween 40. 100 ml of oxygen-saturated distilled water is added when the chloroform is evaporated under vacuum and stirred vigorously. 4 ml of this mixture is added to test tubes containing different concentrations of the sample. The zero time of the absorbance point at 470 nm is immediately measured in each tube. The samples are incubated at 50°C for 2 hours. Quercetin, α-tocopherol and BHT are utilized as standards. A free blank of β-carotene is utilized.

Oxidation of low-density lipoproteins - LDL. From fresh blood samples LDL is isolated and oxidation is initiated with Cu (II) or APPH. In order to control the lipid hydroperoxide production the peroxide concentration is measured, and the peroxidation of lipid components is monitored at 234nm to control the production of conjugated dienes.

Methods based on Simple Electron Transfer - SET reaction mechanisms include: Iron Reducing Antioxidant Power - FRAP. It is a test utilized to measure cells or tissues redox state because the reaction detects compounds with redox potentials of <0.7 V, which is the redox potential of Fe^{3+}-TPTZ. A colored product is measured after 2,4,6-trifiridyl-s-triazine (TPTZ) ferric reduction. Hydroxylation and conjugation degree in polyphenols is related to the reducing power, therefore, FRAP cannot act by trapping radicals, especially thiols and proteins.

Methods exist that utilize both mechanisms: HAT and SET. Among these is the Throlox Equivalent Antioxidant Capacity - TEAC method, which involves $ABTS^{+}$ that is a long-lived chromophore cation that is intensely colored because of ABTS oxidation by peroxyl radicals. AOC measures the ability of antioxidants reduce the color intensity of $ABTS^{+}$ radical. The results are expressed relative to Trolox [49].

The $ABTS^{+}$ method, developed by Rice-Evans and Miller in 1994 and modified by Re *et al.*, 1999, is one of the most widely used antioxidant methods in plant samples, together with the DPHH method. This is a decolorization test which measures the antioxidant capacity in lipophilic and hydrophobic samples, where the antioxidant concentration and absorption time of the radical cation are important. The oxidation reaction of ABTS by potassium persulfate produces blue/green $ABTS^{+}$ chromophore, which in the presence of hydrogen donor antioxidants is measured spectrophotometrically at 734 nm. Trolox is an analog of vitamin E, which is employed as a positive control. Results are expressed as Trolox Equivalent Antioxidant Capacity (TEAC/mg).

In the TEAC assay, Muzolf *et al.*, 2008 [50] found that catechin's antioxidant capacity increases with the pH increase in the culture medium. Catechins with galloyl group TEAC values are higher than quercetin TEAC values and still much higher than cyanidine values in the pH ranges utilized. There are other modifications to the TEAC test: TEAC I allow the evaluation of hydrophilic antioxidants. Carotenoids, tocopherols, and other lipophilic antioxidants can be measured using TEAC II, while TEAC III can be applied to both [51].

The 1,1-diphenyl-2-picrylhydrazine (DPHH) method is also employed to evaluate the antioxidant capacity of plant samples. This method is based on capturing the DPPH free radical by addition of a radical or antioxidant species that induces decolorization of the DPHH solution. This radical is stable and reacts with compounds that can donate hydrogen atoms. Antioxidant activity is measured by the decrease in absorption at 515 nm. Solutions of DPHH (0.1mM) are prepared in methanol and 4 ml of this solution is added to 1 ml of sample solution in methanol at various concentrations. The absorbance at 517 nm is measured thirty minutes later. When there is an absorbance reduction in the reaction mixture, it means a high antioxidant activity of the tested compounds. Through this technique it has been demonstrated that the presence of $3-OH$ groups in the flavonoid molecule gives it a high antioxidant capacity [26].

To determine the total phenolic content, Folin-Ciocalteu's reagent is employed. A solution (0.5 ml) of Folin-Ciocalteu reagent diluted with 7 ml of deionized water, is mixed with 0.2 ml of sample. Before adding 0.2 ml of sodium carbonate solution, the solution containing Folin-Ciocalteu reagent is standardized at 25 °C for 3 minutes. The resulting mixture can rest for 120 min before measuring the absorbance at 725 nm. Gallic acid is employed as the calibration curve standard. The results are expressed as gallic acid equivalents (GAE) per liter of sample (mM/l).

The antioxidant capacity of a compound on the superoxide O^{-}_{2} radical can be measured by the PCL test for both the aqueous phase (ACW) and the lipid phase (ACL). It allows the measurement of the antioxidant capacity of pure substances or matrix components having hydrophilic or hydrophobic nature and coming from plants, humans, or synthetic material. The PCL method is very sensitive and fast. An acceleration of about 1000 times the *in vitro* oxidation reaction can be obtained in the presence of a suitable photosensitizer.

NBT is another assay employed for measuring the antioxidant capacity on O^-_2 superoxide radicals, described in 1971 by Beauchamp and Fridovich. The superoxide radical scavenging capacity is analyzed with the hypoxanthine /xanthine system, coupled to the reduction of nitro blue tetrazolium-NBT which is measured with spectrophotometer. The reaction mixture contains a series of buffer solutions: 125 µl of buffer (50 mM KH_2PO_4/KOH pH 7.4), 30 µl of a 3 mM hypoxanthine solution, 20 µl of a 15 mM Na_2EDTA solution, 50 µl of a 0.6 mM NBT solution, 50 µl of xanthine oxidase (1 unit per 10 ml buffer) and 25 µl of plant extract (10 µg sonicated solution/ 250 µl buffer). After the addition of xanthine oxidase, the microplates are read at 540 nm. The trapping activity on superoxide is expressed as percentage inhibition, and for target the buffer is employed instead of the extract.

Total Flavonoid Content-TFC is a method that allows the measurement of flavonoid extract concentration based on the method described by Moreno *et al.*, 2000. 1 ml of methanol solution containing 1 mg of extract is added to test tubes containing 0.1 ml of 10% $Al(NO_3)_3$, 0.1 ml of 1M CH_3CO_2K and 3.8 ml of CH_3OH. The absorbance at 415 nm is measured after 45 min at room temperature. Quercetin is employed as standard, and the results are expressed as quercetin equivalents.

7. SOME TECHNIQUES UTILIZED TO DETERMINE THE BIOAVAILABILITY OF FLAVONOIDS

In epidemiological studies, the consumption of nutritional and non-nutritional components present in the diet is usually carried out considering nutritional measures such as dietary history, food frequency questionnaire and diet diaries. Food composition data are transferred to databases of specific food components, nutritional or non-nutritional. For flavonoid and therefore polyphenol research, this methodology is imprecise as these databases have very limited information on polyphenol concentration and diversity in plant foods.

Several factors are considered in epidemiological studies, such as: the source of the flavonoid (from natural extracts, market products or chemically pure substances); the design and study population; the methodology for collecting blood or urine samples; and the analysis of flavonoids in plasma and urine. Blood and urine content of phenolic compounds in humans can be measured by the following techniques:

Hydrolysis: The flavonoids in foods are usually glycosylated. The hydrolysis acid or enzyme is frequently used to simplify the analytical procedure and the respective aglycones and free acids are consequently detected and quantified. Because flavonoids are found in methylated, sulfated and glucuronidated forms in humans, enzymatic hydrolysis by means of fulfatasas and β-glucuronidasas is utilized.

Cleaning procedures; The solid phase extraction column (SPE) is utilized to remove matrix components that may alter the analysis of human plasma, serum, or urine samples. However, some techniques such as immunoassays may require only sample preparation.

Separation and detection systems: High-performance HPLC liquid chromatography and gas chromatography and GM-MS mass spectrometry are the two main separation techniques used for the quantification of polyphenols, although it has increased the utilization of liquid chromatography and LC-MS/MS mass spectrometry. For example, this method allowed the detection and quantification of concentration limits from 1.1 to 2.6 nmol/L and from 3.8 to 8.7 nmol/L in plasma and from 0.8 to 1.8 nmol/L and from 2.6 to 6.0 nmol/L in urine for catechin metabolites after consumption of green tea extracts [52].

For flavonoids, however, HPLC is the method of choice, followed by coupled detection systems that include diode network detectors, mass selective detectors, as well as chemical and fluorometric detectors, as demonstrated by studies on the detection of catechin quantities in human plasma atomoles [53]. These compounds are identified by mass fragmentation; however, a simple selective mass detector may not achieve the necessary sensitivity.

Electronic Paramagnetic Resonance (EPR) or Electron Spin Resonance (ESR) is a sensitive spectroscopic technique for unpaired electrons. In the case of an organic molecule, it is a free radical, or a transition metal ion, in the case of an organic compound. Since most stable molecules have matched spin levels, this technique has less application than nuclear magnetic resonance (NMR). Using this technique, some semiquinone radicals formed, the redox properties of polyphenol GTPs and the antioxidant properties of some flavonoids have been identified.

Nuclear magnetic resonance (NMR) spectroscopy is a technique mainly utilized in molecular structure determination, although it can also be used for quantitative purposes. Some atomic nuclei subjected to an external magnetic field absorb electromagnetic radiation in the radio frequency region. As the exact frequency of this absorption is environmentally dependent on these nuclei, it is then employed to determine the molecule structure where they are found. The monodimensional spectrum of ^1H (^1H NMR) provides information about the number and type of different hydrogens in the molecule. The position in the spectrum (chemical displacement) determines the chemical environment of the nucleus, and therefore gives information about functional groups to which they belong, or which are close. Together with MS/MS mass spectrometry and ESR, this technique has been employed for the analysis of GSH products formed when peroxidase/H_2O_2 has oxidized flavonoids [54].

For the analysis of plasma and urine flavonoids, the following pharmacokinetic parameters have been established: AUC area under plasma concentration curve with respect to time; CL/F apparent renal clarity; CL_R, renal clarity; C_{max}, maximum plasma concentration; T_{max}, time to C_{max}; K_a and K_e of absorption and elimination rate constants; %UR, urine recovery of bioactive compound; Vc/F, volume of apparent distribution; $T_{1/2}$ half-life time in the body or time for 50% flavonoid removal. The respective data analysis is also performed [55].

Mass spectrometry is a technique characterized by its high sensitivity and specificity. It is utilized to analyze biological samples mixtures and is also employed in combination with other chromatographic techniques.

Direct analysis of volatile compounds is possible in gas chromatography-mass spectrometry (GC-MS). In this technique, the separation properties of GC are combined with the sensitivity and selectivity of electron impact ionization (EI) mass spectrometry [56]. This latter technique depends on the ion source and ionization mode. Thus, at the EI interface, the electron beam interacts with the molecules of the analyzed compound, increasing its internal energy and producing a mixture of complex ions. Such ions can be fragmenting ions or molecular ions and their relative abundance can be utilized for compound identification. Molecular ions are often absent in the mass spectrum due to excessive fragmentation. Therefore, this ionization mode is less applicable for complex mixtures analysis.

By selected ion monitoring (SIM), a simple chromatogram of the ions of interest can be obtained. In SIM, the mass spectrometer is adjusted to scan a very small mass range (often 1 amu or less), so compounds with selected masses are detected and plotted.

By selected ion monitoring (SIM), a simple chromatogram of the ions of interest can be obtained. In SIM, very small mass range (often 1 amu or less) can be scanned with adjustments made to the mass spectrophotometer, so mass-selected compounds are detected and plotted.

Flavonoid glycosides have been found to exhibit homolytic and heterolytic cleavage induced by collision of the *O*-glycoside bond producing deprotonated aglycone radicals and aglycone-produced ions. Quercetin (quercetin -3-*O*-

ramnoside) may produce a stable aglycon radical anion after collision-induced dissociation of the precursor [M-H]⁻. Quercetin could produce fragment radicals like the antioxidant property of a hydrogen atom or electron donation from the flavonoid.

Flavonoids are substrates for enzymes such as catech *O*-methyltransferase in the small intestine, phase I and phase II enzymes in the liver and other tissues, as well as for UDP-glucuronosyl transferase and *β*-glucosidase. Furthermore, the flavonoids metabolism involves hydrolysis and degradation processes in the colon due to microbial enzyme catalysis.

Some catechins (epicatechin and catechin) in the small intestine are methylated and glucuronidated. After consumption, plasma levels of catechin and its metabolites 3'-*O*-methylcatechin (3'MC) were determined by GC-MS. Following enzymatic hydrolysis, the glucuronide and sulfate conjugates were determined.

The selectivity and sensitivity in the extraction, separation, and quantification process of five catechins in human plasma was greatly improved when the tandem LC-MS system was used. The use of this system has allowed us to understand the bioavailability and biotransformation of EGCG. Following green tea consumption, levels of 4'4'-di-*O*-methyl-EGCG (4'4"-diMeEGCG) were detected in plasma and urine. The same as (-)-5-(3',4'-dihydroxyphenyl)-γ-valerolactone and (-)-5-(3',4',5'-trihydroxyphenyl)-γ-valerolactone and the along with other possible ring fusion metabolites such as (-)-5-(3',5'-dihydroxyphen-l)-γ-valerolactone.

Metabolism of protocyanidins occurs in the stomach or small intestine. However, it has been observed that they are stable at stomach pH. In turn, several low molecular weight metabolites are formed by bacterial action in the colon. By the LC-MS-MS method, it has been possible to identify 14 metabolites of aromatic acids in samples collected in urine 24 hours after consumption.

For the analysis of intact sulfate and glucuronide conjugates of isoflavone, the LC-MS-MS system has also been applied. Some flavonoid metabolites that have been identified from biological fluids are presented in the following table [56].

Table 3. *In vivo* metabolites of some flavonoids.

Parental Compound	Metabolite	[M-H]⁻
Genistein	Dihydrogenistein	271
	Tetrahydrogenistein	273
	6′- hydroxy-*O*-desmethylangolensin	273
	4-Ethylphenol	121
Daidzein	Dihydrodaidzein	255
	Tetrahydrodaidzein	257
	Equol	241
	Cis-4-OH Equol	257
	3′,7-Dihydroxyisoflavan	241
	O-Desmethylangiolensin	257
Quercetin	3,4-dihydroxyphenylacetic acid	167
	m-hydroxyphenylacétic acid	151
	4-Hydroxy-3-methoxyphenyl acetic acid	181
EGCG	-3′-*O*- y 4′-*O*-methyl EGCG	471
	-3″-*O*- y 4″-*O*-methyl EGCG	471
	-4′- 4′-Di-*O*- methyl EGCG	485
	(-)-5-(3′,4′,5′-trihydroxyphenyl)-γ-valerolactone	223
	(-)-5-(3′,4′-dihydroxyphenyl)-γ-valerolactone	207
	(-)-5-(3′,5′-dihydroxyphenyl)-γ-valerolactone	207

8. REACTIVE OXYGEN SPECIES MEASUREMENT

Reactive oxygen species and their derivatives can be measured utilizing several techniques that may vary in sensitivity and safety. Among the main techniques are those to measure the presence of O_2^-, ·OH, H_2O_2, fluorescence techniques to measure the redox state and electron paramagnetic resonance spectroscopy techniques, based on spin trapping or direct electron paramagnetic resonance spectroscopy (EPR). The presence of free radicals can only be directly detected by the EPR technique.

Spin trapping compounds such as DMPO or DEMPO are combined with EPR to optimize the efficiency of the EPR technique, allowing the measurement of O_2^- production in cells, and the *in vivo* or *ex vivo* production of O_2^- or ·OH. When cells and tissues are utilized, the presence of radicals may not be detected because spin adducts can become silent EPR products. Alternatively, techniques such as MS, HPLC, high-field nuclear magnetic resonance and antibody-immune based detection can be utilized for the detection of spin or its metabolites. The utilization of hydroxylamine combined with an appropriate SOD control is also a useful technique for the identification of the superoxide radical $O_2^{.-}$.

The cytochrome c reduction technique is recommended in systems free of reducing equivalents or enzymes but is not recommended in cells and tissues. Nitroblue tetrazolium technique is utilized for measurement of ROS generation in cells *in vitro*. Due to its high sensitivity, the DHE dihydroethidium fluorescence technique combined with appropriate SOD controls is useful for imaging and measuring the presence of $O_2^{.-}$ intracellular and in tissues.

In preclinical studies, it has been shown that oxidative stress caused by reactive oxygen species has an important effect on vascular physiology and pathology. However, the measurement of these molecules in cells and tissues is difficult due to the high reactivity of ROS. Measurement of end products or biomarkers of oxidative stress are some of the indirect methods employed to measure oxidative status in clinical studies. In humans the identification of oxidative stress has been possible with the use of circulating biomarkers.

In clinical studies, the presence in blood and urine of the products of lipid liperoxidation are frequently utilized to evaluate oxidative stress. Among these products are lipid hydroperoxides, whose high plasma levels have been found in patients with diseases caused by high oxidative stress such as peripheral artery disease, heart failure, ischemic heart disease and type I diabetes mellitus, among others.

Biomarkers employed to measure oxidative stress include the measurement of TBARS (thiobarbituric acid reactive substances) in plasma, F_2-IsoPs isoprostanes, HNE 4-Hydroxy-Trans-2-Nonenal and TAC-total antioxidant capacity, among others.

A measure of the antioxidant effect of non-enzymatic defense in biological fluids is the total non-enzymatic antioxidant capacity (TAC). It measures low molecular weight water-soluble and fat-soluble antioxidants, such as vitamin C, vitamin E, bilirubin, thiols, flavonoids, carotenoids, and urate. In humans the TAC test has often been employed to evaluate the effect of antioxidants in the diet.

However, there are difficulties in measuring changes in oxidative stress after consumption of antioxidant supplements because antioxidant effect is not as marked, for example, intracellular SOD reacts 10000 times faster than ascorbic acid; lack of specificity of antioxidants for example, they are not targeted to subcellular sites such as mitochondria and the pharmacokinetics of antioxidants that does not allow sufficient accumulation in tissues to be detected, among other factors [57].

Jiang *et al.*, 2019 [58], utilized the fluorescent technique DHE dihydroethidium to determine the production of the superoxide radical O_2^-, aminophenyl fluorescein APF to determine the production of the ·OH radical and singlet oxygen sensor green SOSG to determine the production of $^1O_2$, by flavonoids irradiated with UV radiation or X-rays. They found that flavonoids with double bonds at the 2,3 position and a 3-OH produce O_2^-, possibly because the energy absorbed by the 3-OH group and the 2,3 double bond activates the double bond and transfers an electron to O_2 to produce O_2^-, while the 3-OH group is separated from the flavonoid molecule to produce ·OH.

9. INTESTINAL ABSORPTION AND PLASMA LEVELS OF FLAVONOIDS

The glycosides, which are water-soluble and chemically stable, are the form in which flavonoids are found in some plants. Deglycosylation is the first step in flavonoids metabolism once consumed and before absorption, it is carried out by a *β*-glucosidase present in the lumen of the ciliated epithelium of the small intestine. Flavonoids present *O*-methylation of the phenolic rings as well as glucuronidation and sulfation during their passage through the small intestine and liver. Flavonoids can reach the colon *via* the enterohepatic circulation where intestinal microflora perform various modifications including ring cleavage, hydrolysis and dehydroxylation to yield low molecular weight compounds.

Flavonoids present in foods are in glycosylated form, except flavanols. In the metabolic process that glycosides present, although in the stomach many resist acid hydrolysis and reach the intestine intact, however, only aglycones and a few glycosides are absorbed there.

The sugar group is a determining factor for the efficiency of glycoside flavonoids absorption into the intestine. It has been shown that onion quercetin glycosides are better absorbed (52%) than pure aglycone (24%) which suggests that the glycosides can be transported to the enterocyte *via* a glucose transporter, dependent on Na^+ SGLT1 where *β*-glucosidase cytosolic hydrolyzes them [59]. This difference in absorption is also since aglycone is chemically unstable at the pH and temperature conditions of the small intestine, while quercetin glycosides can be very stable. In intestinal cells, some flavonoids inhibit the non-Na^+-dependent facilitated diffusion of monosaccharides, thus Na^+-dependent ATPase transport gains efficiency [60]. Furthermore, plasma concentrations of quercetin metabolites have been shown to have a longer half-life than catechin metabolites.

Absorption is influenced by glycosylation; however, it does not affect surrounding metabolites. Thus, intact glycosides of quercetin, genistein and daidzein were not recovered in plasma or urine after ingestion of these compounds either pure or from food. Miyake *et al.*, 2006 [61], found differences in levels AUC_{0-4h} (µmol h/L) for eriodictyol 5.18, homoeridictyol 1.49, and hesperidin 5.62, after ingestion of glycoside flavonoids, and 21.93, 13.44, and 10.301 respectively, after intake of aglycon flavonoids 1 hour after lemon juice consumption.

However, anthocyanins are an exception as intact glycosides are the main surrounding forms, possibly due to aglycone instability, specificity in the absorption mechanism or anthocyanin metabolism, although glucuronides and anthocyanin sulphates have been found in human urine samples. For quercetin, the glycosylation process that occurs in absorption has more effect than for isoflavonoids. In addition, flavonoid absorption depends on the dose used, the administration vehicle, dietary history, and sexual differences among the study population [30].

Once absorbed, flavonoids initiate phase II biotransformation where methylation, sulfation and glucuronidation reactions occur through bioconjugation processes. Methylation can occur at the C3' or C4' position of the flavonoid molecule. This reaction is catalyzed by catechol-*O*-methyl transferase enzyme that catalyzes the transfer of a methyl group from S-adenosyl-L-methionine to flavonoids such as catechin, quercetin, luteolin and cyanidin. The degree of plasma methylation of catechin is lower than that of quercetin. Although present in many tissues, this enzyme has a higher activity in the liver and kidneys [61]. By increasing its hydrophilicity, these metabolic processes facilitate biliary and urinary detoxification and elimination. However, after ingestion of tea, a significant concentration of 4'-methylepigallocatechin in plasma has been found.

For flavonoids, sulfation reactions occur mainly in the liver. Sulfotransferase enzymes catalyze the transfer of a sulfate group from the 3'-phosphoadenosine-5-phosphosulfate to a hydroxyl group of the flavonoids. Glucuronidation occurs in the intestine and liver and high conjugation rates are observed at the C3 position. The enzymes UDP-glucuronyltransferases catalyze the transfer of glucuronic acid from UDP-glucuronic acid to flavonoids and these are found in the endoplasmic reticulum of some tissues. Quercetin metabolites in addition to sulfation are also glucuronidated, whereas catechin metabolites are only glucuronidated.

Despite efficient conjugation mechanisms, generally after consumption of nutritional doses, free aglycones may be in low plasma concentrations or absent. However, catechin aglycones from green tea may form a significant part of the total plasma amount (> 77% for epicagalotechin gallate) [62].

The circulation of flavonoid metabolites through the bloodstream is by binding to proteins especially serum albumin. It is unclear whether flavonoids need to be bound to albumin to exert their biological activity or whether they do so in free form, although their affinity for albumin depends on the chemical structure of the flavonoid [63].

About 25% of the flavonoid aglycones are absorbed as bile micelles in epithelial cells and pass into the lymph, following release of the aglycone glycosides by bacterial enzymes in the intestine. Flavonoids are transported by the lymph into the portal blood from the liver and about 80% is possibly absorbed. Hepatocytes transport the flavonoids to the Golgi apparatus and peroxisomes where they are oxidatively degraded. Another part is found as conjugates that retain their antioxidant properties.

A plasma concentration of approximately 1μM of intact flavonoids can only be reached with repeated ingestion of flavonoids during a certain study. Maximum concentrations are reached 1-2 hours after ingestion. In a study conducted by Manach *et al.*, 2003 [64], they found that the maximum plasma hesperidin concentration was 139 and 387 μg/l for doses of 0.5 and 1.0 liter and the plasma naringin concentration was 16 and 54 μg/l also for doses 0.5 and 1.0 liters of orange juice. The same authors found in another study that the maximum plasma concentration reaches 0.4-0.6 μM after ingestion of 115 mg of naringin aglycone from grape juice and 0.5-0.6 μM after ingestion of 115 mg of hesperidin aglycone from orange juice.

In pharmacokinetic studies of flavanone agglicones Kanaze *et al.*, 2007 [65], they found that the maximum plasma concentration C_{max} was 825.78 ng/ml for hesperidin. For naringin, 2009.51 ng/ml was obtained after consumption of 135 mg of each compound. After ingestion of 120 mg of decaffeinated green tea, EGCG, EGC and EC plasma concentrations were 46-268, 82-206 and 48-80 μg/l, respectively (Table **4**). Moon *et al.*, 2008 [55], in their study of quercetin pharmacokinetics in humans, found for quercetin aglycone, the maximum plasma concentration Cmax, was 15.4 ng/ml, while C_{max} for quercetin conjugate metabolites was 336 ng/ml after consumption of pure quercetin.

Flavonoids enter the tissues especially in the intestine and liver where they are metabolized.

It has been proven that among catechins, the ECG has a great capacity to cross cell membranes and the EGCG can cross the hematoencephalic barrier, which favors its activity as a neuroprotector. The rest of the flavonoids are secreted in the feces and some in the urine. Considering that flavonoids do not accumulate in the liver, the products are transported into the bloodstream by organic acids such as cinnamic acid, caffeic acid and their derivatives and excreted in the urine [62]. Many authors consider the half-life of a flavonoid in the human body to be 3 to 6 hours.

Table 4. Bioavailability of some flavonoids or foods containing flavonoids.

Flavonoid	Source	Amount of Flavonoid Consumed (mg)	Maximum Plasma Concentration (μg/l)	Urinary Excretion (% of Consumption)	Reference
Flavanols	-	-	-	-	-
EGCG	Tea	88	46-268	87	[68]
EGC	Tea	82	82-206	92	[68]
EC	Tea	32	48-80	95	[68]
Flavanones	-	-	-	-	-
Hesperidin	Orange juice	444	139-387	4.1-6.4	[69]
Hesperidin	Orange juice	51-102	43.4	-------	[70]
Hesperidin	Orange juice	6-12	79.8	-------	[70]
Hesperidin	Capsule	135	825.78	3.26	[65]
Narinjin	Orange juice	22.6-45	16-54	7.1-7.8	[69]
Narinjin	Orange juice	51-102	16.4	------	[70]
Narinjin	Orange juice	6-12	34.0	------	[70]
Narinjin	Capsule	135	2009.51	5.81	[65]
Flavonols	-	-	-	-	-
Quercetin aglycone	Pure composite capsule	500	15.39	0.05-3.6	[55]
Quercetin metabolites	Pure composite capsule	500	447.8	0.08-2.6	[55]
Isoflavanones	-	-	-	-	-
Genistein	Soybean extract capsule	46.12	254.6	51.4	[67]
Genistein	Soybean extract capsule	40.27	261.9	33.2	[67]
Daidzein	Soybean extract capsule	46.12	508.1	65	[67]
Daidzein	Soybean extract capsule	40.27	567.7	65	[67]

Flavonoids can also be eliminated in the bile, especially the largely conjugated metabolites, while small conjugates such as monosulfates are eliminated in the urine. From the maximum plasma concentration depends on the total amount of metabolites excreted by the kidney. However, for flavones obtained from citrus fruits, the excretion percentage is high since it is between 4 and 30% of the consumed [66], for isoflavones it is between 16 and 65% for daidzein, 10 and 33.2% for genistein [67]; while 0.3-1.4% of the consumed dose of flavonols such as quercetin and its glycosides were obtained (Table **4**).

The metabolism of quercetin and catechin metabolites causes the differences in the bioavailability of each and affects their solubility in the organic fluids, being responsible of the different pathways of elimination of both flavonoids. Catechin is eliminated mainly by urine while quercetin is eliminated by bile [68].

Other recent studies have reported a urinary excretion of 0.05-3.6% for quercetin aglycone and 0.08-2.6% for quercetin conjugates (Table **4**). On the other hand, for flavanols such as tea catechins the urinary recovery was 0.5-6%, for wine catechins it was 2-10% and for cocoa catechins it was more than 30%. While in other studies the recovery of EGCC, EGC and EC from urine samples after tea consumption were: 87, 92 and 95%, respectively. For other flavonoids such as anthocyanins these percentages are very low, 0.005-0.1% of the total consumed.

Furthermore, flavonoids that are not absorbed in the intestine reach the colon, where the microflora hydrolyze the glycosides to aglycones and metabolize them to various aromatic acids. The intestinal microflora also affects the metabolism of the isoflavonoid glycosides, as they are hydrolyzed to aglycones or transformed to equol from daizein, which are active metabolites. It is considered that only 20-35% of the adult population can convert daidzein to equol after consumption of soy or soy derivatives. Vergne *et al.*, 2007 [67], showed that daidzein excretion was significantly lower in equol producers compared to non-equol producers in the total time of soy elimination.

The flavonoids have a short retention time in the intestine, low water solubility and a low absorption coefficient, therefore flavonoids in humans cannot cause acute toxic effects unless produced by an allergy to these compounds. However, the intravenous administration of unpurified flavonoid extracts or the consumption of various types of propolis may be toxic, since in some tropical regions highly toxic flavonoids have been found. For example, carrying out toxicology experiments to determine the safety limit of a certain compound, it has been observed that after three weeks of constant consumption of high flavonoid concentrations, hepatocytes can change shape, lead to necrosis and death of the animal [60].

After consumption of foods rich in flavonoids such as flavan-3-ol and procyanidin, studies on the real plasma concentrations of these compounds have limitations. Concentrations found in human plasma after an actual consumption of flavonoids are in the nanomolar interval. Compared to other compounds with similar antioxidant properties, this low availability is kinetically unfavorable as in tocopherol and ascorbate which are present in the bloodstream at micromolar concentrations.

Therefore, after consumption, the antioxidant effect is exerted directly in the bloodstream and other tissues. Thus, the changes observed because of cellular, or tissue oxidation can be explained by other mechanisms that are compatible with the physiological levels reached by polyphenols [71].

The presence of *O*-catechol group (3,4'−OH) in the B-ring is determinant for high antioxidant activity, as was demonstrated by Justino *et al.* in 2004 [72], when they evaluated the structure-antioxidant activity relationship with quercetin and its metabolites. They found that metabolites with substitutions at the 3'−OH and 4'−OH groups have lower antioxidant activity than quercetin. Meaning that, after metabolic transformation, quercetin's antioxidant activity may decrease if a conjugation reaction occurs at one of the *O*-hydroxyl groups.

Flavonoids are widely metabolized during absorption generating circulating glucuronides, sulphated conjugates, *O*-methylated forms, and small phenolic acids. In the study realized by Pollard *et al.*, in 2006 [73], they found that epicatechin and quercetin flavonoids that contain catechol group were efficient inhibitors of tyrosine nitration induced by peroxynitrite. In general, *O*-methylation of the B-ring and glucuronidation of the A-ring decreased the potential of these flavonoids to inhibit tyrosine nitration.

The quercetin agglicone flavonol, has a 3',4', -dihydroxy in the B ring, saturation between C2 and C3 and a carbonyl motif in position 4, all this structure is strongly involved in its antioxidant activity. The reaction of ONOO⁻ with quercetin results in the oxidation and formation of methyl-quinone and *O*-quinone that react with GSH thiol to produce glutathione adducts in the A and B rings. Quercetin quinone's reaction with GSH *in vivo*, *e.g.* at inflammation sites where ONOO⁻ is formed, may reduce the potential damage induced by quinone species. The quercetin B-ring *O*-methylation reduces its ability to trap peroxynitrite and its antioxidant potential. The decrease in the ability of 4'-*O*-methyl metabolite is mainly related to the instability of this metabolite to form methyl quinone species. Quercetin glucuronidation slightly decreases its antioxidant activity, but greatly decreases its ability to trap peroxynitrite [72].

After oral intake of some flavonols such as (-)-epicatechin, (+)-epicatechin, (-)-

catechin, (+)-catechin, (+)-catechin steroisomers, Ottaviani *et al.*, 2011 [74], evaluated the incidence on absorption, metabolism, and urinary excretion of compound chemical properties such as stereochemical configuration and found that flavanol steroisomer absorption and subsequent appearance in plasma occurred from oral ingestion of the individual steroisomers.

The respective steroisomers of the four ingested flavanols were present in the circulation and at the detection limit their corresponding C-2 epimers. Following 2-4 hours after consumption the plasma levels of the four steroisomers had the following order (-)-epicatechin > (+)-catechin = (+)-epicatechin > (-)-catechin, although the ingested amounts of each were equal to 1.5 mg/kg body weight.

The total amount of (-)-catechin metabolites in plasma after 2 h of postingestion was 149±18 nM, while it was 889±114 nM for (-)-epicatechin. After 2 h of consumption the interindividual variation of flavanol metabolites in plasma was 28% (+)-catechin, 32(-)-catechin, 34% (-)epicatechin and 37(+)-epicatechin.They also observed that the total amount of flavanol metabolites in plasma was higher at 2 h than at 4 h after consumption (p<0.05) for each of the four ingested compounds independently of the flavanols' stereochemical characteristics.

Analysis of the urine studies showed that stereochemistry of the flavanols consumed also affected the results in a similar way as in plasma. Thus, (-)-catechin was the steroisomer with the lowest level (4.0±0.4 μmol) found in urine 24 h (p<0.05) after consumption, corresponding to 0.9±0.1% of the amount consumed. The rest of the flavanol stereoisomers have the following levels expressed as micromoles or as a percentage of the amount ingested: (+)-catechin, 9.8±1.4μmol/2.3±0.3%; (+)-epicatechin,10±1μmol/2.4 ± 0.3% and (-)-epicatechin, 13±2 μmol/3.0 ± 0.5%.

Renal excretion of consumed flavanol stereoisomers occurs rapidly after ingestion, with 90 ±5, 94±7, 95±2, and 97±3% of (-)-epicatechin, (-)- catechin, (+)-epicatechin, and (+)-catechin metabolites, respectively, being excreted within the initial 8 hours after consumption.

Chiral chromatographic analysis of the HPLC fractions of catechin and epicatechin obtained from the urine samples determined that the excreted flavanols have the same stereochemical configuration as the ingested flavanols, indicating that *In vivo*, the flavanols are not influenced by stereoisomeric interconversion.

In humans, a main metabolic pathway for flavanols involves the catechol-*O*-methyl transferase (COMT)-dependent pathway. The amounts of 3'-*O*-methylated, 4'-*O*-methylated and unmethylated flavanols were measured in plasma and urine.

3'- and 4'-*O*-methylated derivatives of epicatechin were found in plasma by oral consumption of (+)- and (-)-epicatechin, whereas only 3'-*O*-methylated derivative was found after consumption of (+)- and (-)-catechin. In urine analyses for these metabolites, the 4'-*O*-methylated derivative of catechin was also not found. Regarding plasma levels of the 3'-*O*-methylated metabolites for (-)-epicatechin and (+)-epicatechin were found to be (219 ± 20 nM *vs*. 239 ± 36 nM) and of the 4'-*O*-methylated metabolite were (50 ± 6nM *vs*. 69 ± 15nM) respectively.

However, plasma levels of non-methylated metabolites of (-)-epicatechin were 620 ± 99 nM *vs*. 278 ± 39 nM of (+)-epicatechin $p < 0.05$ (n=7); so that non-methylated metabolites of (-)-epicatechin predominate over its enantiomer (+)-epicatechin, 68 ± 4% *vs*. 47 ± 3%; ($p < 0.05$). In the analysis performed in urine, the non-methylated metabolites corresponded to 75 ± 2% for (-)-epicatechin and 42 ± 2% ($p < 0.05$) for (+)-epicatechin, respectively from metabolites excreted in urine [74].

10. FLAVONOIDS PROOXIDANT ACTIVITY MECHANISM

Scientific evidence is sufficient to demonstrate the capacity of flavonoids as antioxidants, however, necessary evidence also exists to demonstrate their prooxidant activity. Flavonoids' prooxidant activity is directly related to the presence of hydroxyl groups. Thus, in flavonols, the presence of a different number of −OH groups in the B-ring contributes to their antioxidant action as well as to their toxicity. Thus, the same stability of the flavonoid's phenoxyl radical can give rise to a prooxidant action.

Instead of completing the chain reaction (reaction 27) the phenoxyl radical can react with oxygen to produce quinones and superoxide anion, a reaction that occurs in the presence of high levels of metal ions and leads to a prooxidant action of flavonoids.

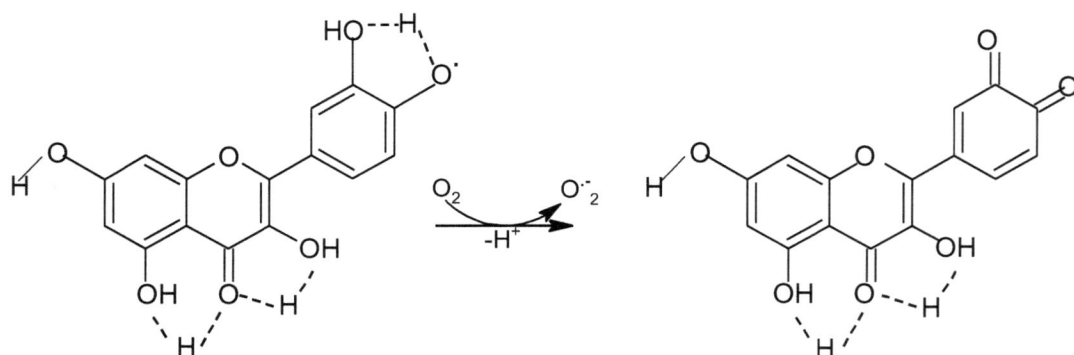

Reaction (27), Flavonoids prooxidant activity.

A flavonoid compound can behave as an antioxidant or a prooxidant. The antioxidant and prooxidant activity of several flavonoids has been evaluated and it has been found that only flavonoids that have an *ortho*-trihydroxyl group on the B- or A-ring has prooxidant activity.

Furthermore, other studies have demonstrated that the presence of the catechol structure in the A- or B- rings confers the flavonoid the property to be pro-oxidant. For this reason, reactive oxygen species (ROS) originate when the catechol group is oxidized by the copper (II) ion. This occurs through the electron transfer-SET mechanism where Fe (III) acquires an electron from the carbon neighboring the 3-hydroxy-4-keto group, originating the iron(II) complex (reaction 28) that produces reactive oxygen species-ROS.

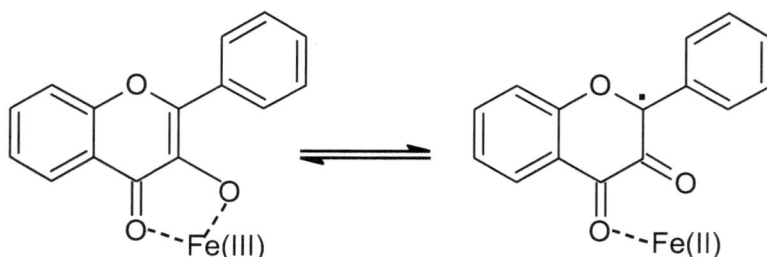

Reaction (28), Pathway for iron complex formation from flavonoid-Fe(III) complex.

Flavonoids may be toxic because of the antioxidant effect they contain, and this may be applicable to other antioxidant free radical trappers [75]. For determining the oxidation products, it is important to know the redox potential of flavonoids.

Quercetin (3,5,7,3',4'-pentahydroxyflavone), has three structural determinants that are indispensable for free radical trapping and/or the flavonoids' antioxidant potential: the 2-3 double bond in conjugation with 4-*oxo* from the carbonyl in the C-ring, a catechol moiety in the B-ring and the presence of 3- and 5−OH groups in the A-ring [76].

Quercetin's action as an oxidant or prooxidant depends on the concentration and source of free radicals. As an antioxidant, quercetin is known to protect cellular components from oxidative damage, reducing cell death caused by the decrease of cellular antioxidants and due to its ability to interact with the lipid bilayer it can modify membrane-dependent processes. Quercetin is also known to have prooxidant activity in the presence of transition metals. However, quercetin also acts directly as a mutagen.

Superoxide O_2^- radical oxidizes quercetin to produce semiquinone radical and H_2O_2. Quercetins in chemical systems can quickly autooxidize to free radicals $\cdot OH$ and semiquinone.

It may be affirmed that there is a toxic effect as a result of free radical metabolism when the following considerations are demonstrated: firstly, free radical formation, secondly, toxicity characteristics that may be consistent with enzymatic or non-enzymatic knowledge of the reactions of free radicals and thirdly, that the concentrations and half-life of the same in the microenvironment of the modified biological targets, make it possible to quantitatively rationalize the effects observed on them [76].

Quercetin oxidation results in the formation of a free radical intermediate *o*-semiquinone (Q^-) through a reaction catalyzed by lactic peroxidase in a typical peroxidase mechanism:

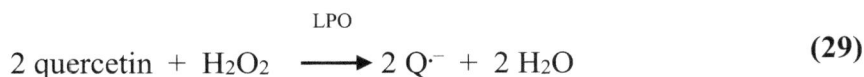

$$2 \text{ quercetin } + \text{ H}_2\text{O}_2 \xrightarrow{\text{LPO}} 2 \text{ Q}^- + 2 \text{ H}_2\text{O} \tag{29}$$

The reaction is followed by disproportionate reaction resulting in *o*-quinone (Q) and parental molecules:

$$2 \text{ Q}^- \longrightarrow \text{ Q } + \text{ quercetin} \tag{30}$$

The reaction 30 has a 6.0 x 107 $M^{-1}s^{-1}$ velocity constant (see also table 5). *O*-semiquinone (Q^-) can also react with O_2 to form O_2^- superoxide and restore Q:

$$Q^- + O_2 \longrightarrow Q + O_2^- \tag{31}$$

This reaction can initiate redox cycling, which depends on the nature and concentration of the reactants and products [76], since *e.g.*, Q^- semiquinone radical can be in anionic or neutral form. Land & Swallow in 1970 [77], employing pulse radiolysis techniques, determined the ubiquinone reduction rates by the reaction constants of electrons solvated with ubiquinone in acid and alkaline solutions. In the presence of H_2SO_4 10^{-2} M the electrons solvated could be converted to hydrogen atoms and give rise to $\cdot CH_2OH$ radicals which reduce ubiquinone:

$$\cdot CH_2OH + Q \qquad HCHO + QH\cdot \tag{32}$$

$QH\cdot$ is a semiquinone neutral radical that has a velocity constant of 1.4×10^9 $M^{-1}s^{-1}$ [77]. The extinction coefficients of the semiquinone radical can also be obtained for anion in alkaline solutions. With an average extinction coefficient of 3000 $M^{-1}cm^{-1}$, the disappearance rate of the semiquinone radical is 4.8 $\times 107$ $M^{-1}s^{-1}$ (Table 5).

$$2 QH\cdot \longrightarrow Q + QH_2 \tag{33}$$

Therefore, the semiquinone radical is considered to produce more anions than protons:

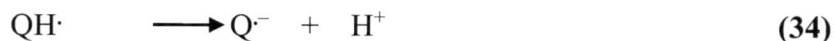

$$QH\cdot \longrightarrow Q^- + H^+ \tag{34}$$

Semiquinone anions are disproportionate and may also react with the neutral radical:

$$Q^- + QH\cdot \longrightarrow Q + QH\cdot \tag{35}$$

From reaction 31 and once the superoxide has formed, it can react with quercetin.

$$\text{Quercetin} + O^-_2 \xrightarrow{2H^+} Q^- + H_2O_2 \tag{36}$$

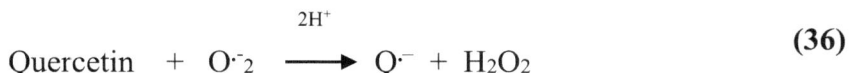

This antioxidant reaction mechanism of quercetin that leads to the formation of reactive oxygen species (prooxidant activity) corresponds to one-electron transfer with concerted proton transfer -SET/HAT in the transient state ($K_4 = 4.7 \times 10^4$ $M^{-1}s^{-1}$).

Table 5. Reaction velocities determined by pulse radiolysis.

Reaction	Reaction constants	Reference
$2 Q^- \rightarrow Q + $ quercetin	$6.0 \times 10^7 \ M^{-1}s^{-1}$	[78]
$\cdot CH_2OH + Q \rightarrow HCHO + QH\cdot$	$1.4 \times 10^9 \ M^{-1}s^{-1}$	[77]
$2 QH\cdot \rightarrow Q + QH_2$	$4.8 \times 10^7 \ M^{-1}s^{-1}.$	[77]
quercitin $+ O^-_2 \rightarrow Q^- + H_2O_2$	$4.7 \times 10^4 \ M^{-1}s^{-1}$	[79]
$PhO\cdot + NADH \rightarrow PhOH + NAD\cdot$	$8.0 \times 10^7 M^{-1}s^{-1}$	[80]
$NAD\cdot + O_2 \rightarrow NAD^+ + O^-_2$	$2.0 \times 10^9 M^{-1}s^{-1}$	[81]
$2H^+ + Q^- + Q^- \rightarrow Q + QH_2$	$5.8 \times 10^7 M^{-1}s^{-1}$	[82]

To yield *o*-semiquinone Q^-, quercetin exhibits electron oxidation, which can be attributed to the formation of O^-_2 (reaction 31) or to the following one:

$$Q^- + O^-_2 + 2H^+ \rightleftharpoons Q + H_2O_2 \tag{37}$$

During quercetin oxidation in the presence of a reducing agent such as GSH, the reaction is affected, which explains the reaction of GS· with Q^- according to reaction:

$$Q^- + GS\cdot \longrightarrow Q + GSH \tag{38}$$

where by the transfer reaction of one electron donated by Q^- it is possible to form GS·

$$GSH + Q^- \rightleftharpoons GS\cdot + \text{quercetin} \tag{39}$$

and/or in the reaction of GSH with O^-_2 such as the following:

$$GSH + O^{\cdot -}_2 + H^+ \longrightarrow GS^\cdot + H_2O_2 \tag{40}$$

The formation of $Q^{\cdot -}$ in an oxidoreductive metabolic cycle of quercetin can lead to pro-oxidant effects when O_2 is present (Fig. **14**).

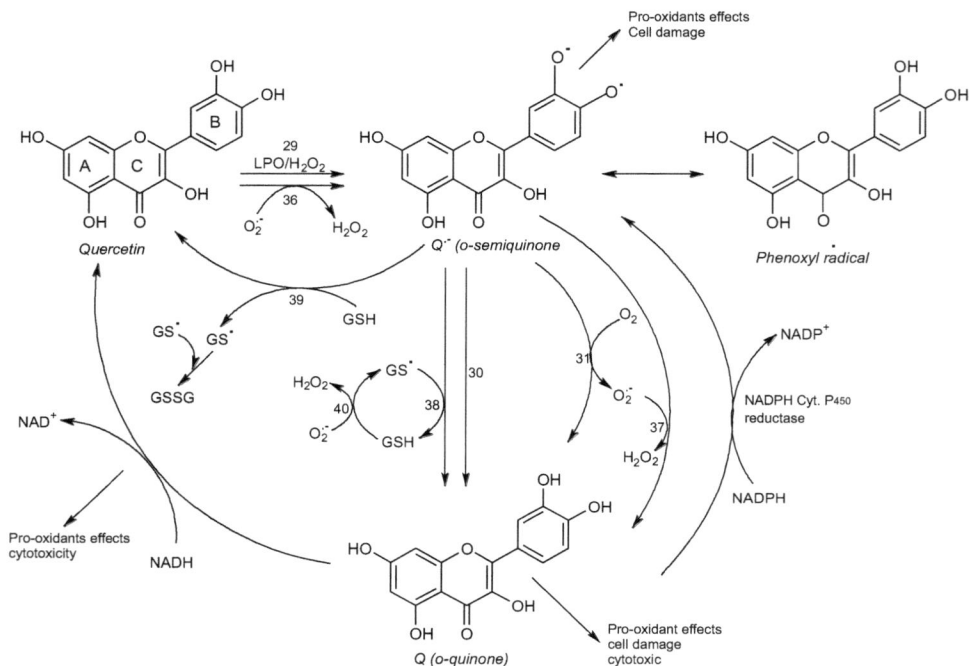

Fig. (14). Summary of oxidoreductive activation of quercetin. The numbers of the reactions are discussed in the text.

Quercetin can express its prooxidant action possibly by $Q^{\cdot -}$ formation and its *o*-quinone molecular product. Such species are toxic and irreversibly bind to various cellular constituents by forming a covalent bond with sulfhydryl groups or other groups, in order to produce free secondary radicals and other products.

In vivo production of small amounts of quercetin $Q^{\cdot -}$ and Q is toxicologically important.

Hodnick *et al.*, 1988 [83], evaluating the electrochemical properties of some flavonoids found $O^{\cdot -}_2$ and H_2O_2 formation as intermediates in the myricetin autooxidation reactions. Myricetin can reduce non-enzymatically oxygen to $O^{\cdot -}_2$ with oxidized myricetin formation. The $O^{\cdot -}_2$ dismute to H_2O_2 and O_2. Oxidized myricetin can accept succinate electrons through the respiratory chain.

Furthermore, it has been demonstrated that the autoxidation of flavonoids requires the presence of hydroxyl groups and only reacts in alkaline solutions, where quinonoid species and small quantities of H_2O_2 are produced.

Some flavonoids are metabolically activated to prooxidants in the absence of autooxidation which is catalyzed by transition metals as demonstrated by Chan *et al.*, 1999 [84]. In NADH oxidation the prooxidant activity of flavones and flavanones is accompanied by formation of oxygen adducts and is related to the redox potential of the phenoxyl radical. Thus naringenin, apigenin and hesperidin have higher redox potentials than other flavonoids. The one-electron redox potential of $NAD\cdot, H^+/NADH$ was 300 mV.

The oxidation of electrons from phenol to phenoxyl radicals is catalyzed by peroxidases that co-oxidize NADH to $NAD\cdot$, and rapidly reduce O_2 to O^-_2 [85] as presented in the following reaction:

$$PhO\cdot \; + \; NADH \; \longrightarrow \; PhOH \; + \; NAD\cdot \tag{41}$$

$$NAD\cdot \; + \; O_2 \; \longrightarrow \; NAD^+ \; + \; O^-_2 \tag{42}$$

With pulse radiolysis techniques, velocity constants of $8.0 \times 10^7 M^{-1}s^{-1}$ [80] have been found for phenol-phenoxyl radical reaction with NADH. The reaction between the phenoxyl radicals and NADH is done by electron direct transfer resulting in $NAD\cdot$ formation. A velocity constant of $2.0 \times 10^9 M^{-1}s^{-1}$ has been reported for the reaction of $NAD\cdot$ with O_2 (Table **5**).

From flavonoids with catechol group on the B-ring, the *o*-semiquinone (Q^-) radicals formed were unstable with rapid kinetic of chemical disproportion and a low probability of reacting with NADH. However, NADH was oxidized by flavonoids with catechol group on the B-ring, but without oxygen intervention, thus indicating that NADH presents two-electron oxidation for the *o*-quinone product. These flavonoids, when activated by $HRP/H_2O_2/NADH$, reduced cytochrome *c* without being affected by superoxide dismutase, suggesting the formation of semiquinone radical and non-radical superoxide. When catechol *o*-semiquinone radicals were stabilized with Zn^{2+} or Mg^{2+}, ascorbate or GSH reacted with the radical.

Reactions 43-45 may explain the NADH-mediated lack of oxygen activation of semiquinone radicals:

$$Q^{\cdot-} + Q^{\cdot-} \longrightarrow Q + catechol \tag{43}$$

$$Q^{\cdot-} + NAD^{\cdot} \rightleftarrows Q + NADH \tag{44}$$

$$Q^{\cdot-} + NAD^{\cdot} \longrightarrow catechol + NAD^{+} \tag{45}$$

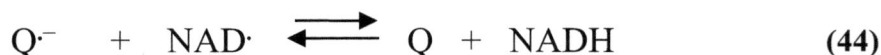

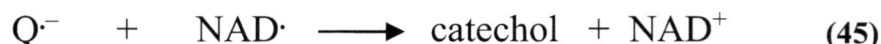

On the other hand, the flavonoid naringenin after forming phenoxyl radicals, in the absence of NADH or other electron donors, presents a series of reactions leading to the oxidative degradation of naringenin including an oxidative action on the C-2 and A ring. However, this naringenin or apigenin oxidation can be prevented by NADH until it has been oxidized, which means that NADH reduces the phenoxyl radicals without the intervention of naringenin or apigenin [84].

In flavanone naringenin and flavone apigenin only the 4'-hydroxy group of the B-ring leads to glutathione GSH oxidation at physiological pH in the presence of H_2O_2 and **peroxidase**. This occurs at catalytic concentrations of both flavonoids. Oxygen conduction was proportional to the oxidation of GSH to GSSG. These results indicate that the phenoxyl radical cooxidizes GSH to form oxygen-activating thiyl radicals. In turn, thiyl radicals initiate reactive oxygen species (ROS) formation.

After oxidation by H_2O_2 and peroxidase to semiquinone radicals the flavonol quercetin and the flavone luteolin reduce GSH without thiyl radical formation through the quercetin-SG and luteolin-SG conjugate, respectively.

The following reactions could explain the results obtained for luteolin (QH_2) oxidation catalyzed by peroxidase/H_2O_2:

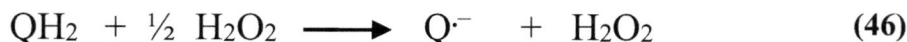

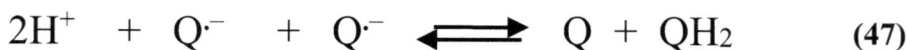

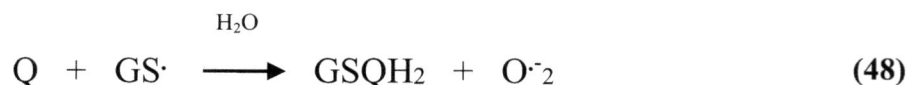

$$QH_2 + \tfrac{1}{2} H_2O_2 \longrightarrow Q^{\cdot-} + H_2O_2 \tag{46}$$

$$2H^{+} + Q^{\cdot-} + Q^{\cdot-} \rightleftarrows Q + QH_2 \tag{47}$$

$$Q + GS^{\cdot} \xrightarrow{H_2O} GSQH_2 + O^{\cdot-}_2 \tag{48}$$

$$GSQH_2 \;+\; Q \longrightarrow GSQ \;+\; QH_2 \qquad\qquad (49)$$

$$GSQ \;+\; GS\cdot \;+\; 2H^+ \longrightarrow (GS)_2QH_2 \qquad (50)$$

The formation of semiquinone radicals (reaction 46) has also been demonstrated for quercetin oxidation catalyzed by lactoperoxidase/H_2O_2 (reaction 29). The disproportionate rate for o-semiquinone quercetin radical is 6.0 x 10^7 $M^{-1}s^{-1}$ (reaction 30) and for o-semiquinone luteolin is 5.8 x $10^7 M^{-1}s^{-1}$ [82] (reaction 47) (Table **5**). The formation rate of the o-quinonoid product is slightly increased by superoxide dismutase (reaction 31); however, electronic transfer catalyzed by oxygen when quercetin has been oxidized by HRP/H_2O_2 did not increase.

Phenoxyl radicals of naringenin or apigenin are more effective in oxidizing GSH than kaempferol phenoxyl radicals or luteolin/quercetin semiquinone radicals because the one-electron redox potentials of naringenin ($E_{p/2}$ = 600 mV) or apigenin ($E_{p/2}$ = > 1000 mV) phenoxyl radicals are more closely related to those of GSH. Whereas the one-electron redox potentials of kaempherol ($E_{p/2}$ = 120 mV), luteolin ($E_{p/2}$ = 180 mV) and quercetin ($E_{p/2}$ = 30 mV) are much lower than those of GSH ($E_{p/2}$ = 850 mV) (51 and 52 reactions).

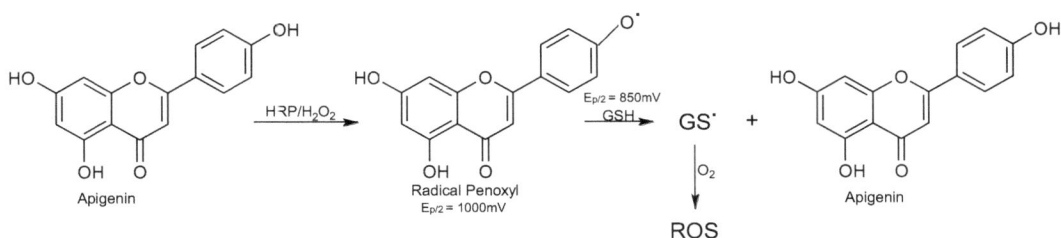

Reaction. **(51)**.

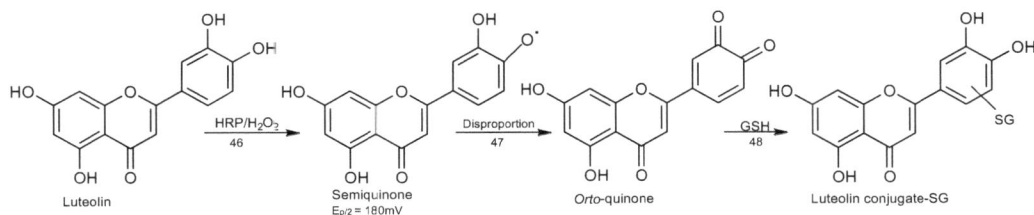

Reaction. **(52)**.

Quercetin does not form B-ring conjugates due to the reaction of GSH with *o*-quinone, unlike luteolin. GSH binds to the 8-position of the A-ring to form an 8-glutathionyl conjugate, as demonstrated by ^{1}H NMR. Similarly, it has been shown that the tyrosinase mechanism involves two-electron oxidation, whereas the peroxidase mechanism involves one-electron oxidation. The rate of reaction of GSH to *o*-quinone of quercetin is higher than the rate of isomerization to form methyl-quinone. However, the electrophilic reactivity of the methyl-quinone isomers is higher than the *o*-quinone of quercetin. Thus, GSH might tend to react more with quercetin methyl quinone, possibly through the Meisenheimer reaction pathway.

From the above observations, Galati *et al.* in 2001 [86], demonstrated that the stoichiometric formation of GSH conjugates without GSH oxidation is possibly due to the low redox potential of quercetin. Peroxidase catalyzes polyphenol oxidation with loss of benzyl hydrogen to produce the phenoxyl radical, which after disproportionation forms methyl quinone, and the formation of methyl quinone from catechols are the two main mechanisms for methyl quinone formation through oxidative metabolism.

Some peroxidases catalyze the one-electron oxidation of catechols to form a semiquinone radical that is converted to an *o*-quinone after disproportionation. Whereas tyrosinases catalyze catechol and/or phenol oxidation to directly form *o*-quinones. When the 4-alkyl substituent of the *o*-quinone has a benzylic hydrogen, then quinones isomerize the methyl quinone. Methyl quinone is quercetin's oxidation product, however, quercetin does not have a benzyl hydrogen so its quinonoid product (methyl quinone) would be formed by a third mechanism involving tautomerization between the 3-hydroxy through the double C2 = C3 conjugate of the *o*-quinone's B-ring.

A semiquinone radical is formed by quercetin oxidation by HRP/H_2O_2. During peroxidative metabolism, quercetin can form glutathione conjugates on the C-2 of the C-ring by quinonoid product formation. Also, the 3-hydroxy group on the C-ring is oxidized due to the disproportionation reaction to a quinonoid product (methyl quinone), which reacts with GSH to form a GS-quercetin conjugate, *via* the Meisenheimer reaction. Thus, quercetin presents an oxidation from its semiquinone radical to the methyl quinone radical in a similar way as kaempferol presents an oxidation from its phenoxyl radical to the quinonoid product (methyl quinone radical).

As mentioned above, due to high redox potentials of naringenin and apigenin, these compounds co-oxidize GSH to GS, producing GSSG, whereas due to the low redox potentials of kaemferol or luteolin/quercetin, GSH was not oxidized. However, from the methyl quinone radicals of quercetin/kaempferol and *o*-quinone radicals of luteolin, GSH conjugates were formed. Galati *et al.* in 2001 [86] then concluded that quercetin/kaempherol was oxidized at the 3'- and 4'-hydroxy groups to form a quinonoid (methyl quinone), whereas luteolin was oxidized to an *o*-quinone.

In addition, pulse radiolysis studies show that ascorbate is more rapidly oxidized by semiquinone radicals containing the catechol ring than *o*-quinone radicals. In catecholamine oxidation by peroxidase/H_2O_2, semiquinone radicals were reduced by ascorbate producing semihydroascorbate radicals, as demonstrated by ESR measurements. In the presence of substrates with catechol rings such as quercetin, catechin or caffeic acid, H_2O_2 detoxification by HRP and ascorbate presented considerably high levels, indicating that a synergistic antioxidant action exists between ascorbate and catechols for intracellular H_2O_2 scavenging [85].

The following reactions may explain ascorbate co-oxidation catalyzed by phenoxyl radicals. The electron oxidation of phenol (PhOH) caused by its reaction with H_2O_2 to form phenoxyl radicals (reaction **53**) is catalyzed by peroxidase, these phenoxyl radicals oxidize ascorbate ($AscH_2$) to form semi-hydroascorbate radicals (reaction **54**), which disproportionate to form dehydroascorbate and ascorbate (reaction **55**):

$$PhOH \; + \; H_2O_2 \; \xrightarrow{\text{Peroxidase}} \; PhO{\cdot} \; + \; H_2O \; + {\cdot}OH \qquad \textbf{(53)}$$

$$PhO{\cdot} \; + \; AscH_2 \; \longrightarrow \; PhOH \; + \; AscH{\cdot} \qquad \textbf{(54)}$$

$$AscH{\cdot} \; + \; AscH{\cdot} \; \longrightarrow \; AscH_2 \; + \; Asc \qquad \textbf{(55)}$$

Unlike ascorbate co-oxidation, catalytic amounts of H_2O_2 and flavonoids are required for NADH co-oxidation in addition to extensive electron transfer through oxygen. Under pH 7 conditions the one-electron reduction potential for $NAD{\cdot},H^+/NADH$ is 282 mV [85] (reactions 41 and 42).

GSH reduction without oxygen-mediated activation or electron transfer is performed by some flavonoids such as quercetin, luteolin and fisetin containing the catechol group on the B-ring. The semiquinone radicals of these flavonoids oxidize GSH at levels below their disproportionation rate because their redox potentials are lower than those of GSH, as indicated above. The methyl quinone products of fisetin and quercetin and the two-electron oxidation products of the *o*-quinone of luteolin react with GSH conjugates.

Ferreira *et al.*, 2009 [87], utilizing spectroscopy techniques - EPR evaluated the free radicals generated in alkaline autooxidation reactions and reactions with O_2^- radicals in several catechins. They found that in EGCG which has two pyrogalol groups, the B-ring is the site of alkaline autooxidation and that oxidation by O_2^- generates an unstable radical in the D-ring; followed by degradation that would lead to radical formation of the relatively stable gallic acid. D-ring oxidation also occurs in the absence of metal chelation, although its importance in B-ring oxidation seems to decrease with increasing pH.

Flavonoids also inhibit oxygenase enzymes, which include many oxidoreductases, because they all have the following catalytic aspects in common: 1) The mechanisms by which they all act is free radicals that are trapped by flavonoids. 2) Although some of them utilize tetrahydrofolate as a catalyst for electron transfer, flavonoids can prevent its participation. 3) The pyridine and flavin nucleotides utilized by some enzymes in their electron chain can be blocked by the action of flavonoids. 4) As an essential component of their catalytic mechanism, oxygenases contain Fe^{2+} and Cu^{2+} and flavonoids have a high affinity for heavy metals. Thus, the oxidation/reduction potentials of these ions are displaced and the positions and architecture of the ligands in the enzymes change, *e.g.*, the stimulatory effects of a flavonoid aglycone may be partially or totally due to its binding to the cofactor [30].

In 2007, Boots and collaborators [88], demonstrated that quercetin reverses the damage caused by H_2O_2 in cells and tissues. However, several oxidation products are formed although highly reactive species are neutralized. For example, the first oxidation product of quercetin may be the semiquinone radical. However, this radical is unstable and rapidly presents a second oxidation that produces quercetina-quinone QQ, which has four tautomeric forms: three methyl quinones and *ortho*-quinone. QQ is highly reactive towards thiols, which show a preference for reaction with GSH. The union of oxidized quercetin with GSH is reversible. At low GSH concentrations, quercetin oxidized may react with thiol-proteins to give rise to a relatively stable protein-quercetin product.

The binding of quercetin oxidized to thiol groups of proteins has been demonstrated in isolated membranes and lymphocytes, as well as in blood plasma and can cause loss of function and damage to enzymes such as calcio-ATPase of the sarco and endoplasmic reticles, which is responsible for intracellular calcium retention. Several studies on quercetin oxidation have demonstrated that it causes an increase in intracellular free calcium levels, a decrease in GSH levels, and a breakdown of LDH. Since GSH efficiently protects against protein arilación and loss of function, GSH reduction increases quercetin toxicity. It has also been found that loss of GSH caused by oxidation of quercetin reactive thiol products such as 4-methyl-*ortho*-benzoquinone cause peroxide-induced cytoskeleton reorganization and clots formation in endothelial cells.

In addition to the above and although there are successful results in the work done with cell cultures to study the effect of flavonoids as antioxidants, it is important to understand that cells in culture are under oxidative stress, which alter their properties in multiple ways even promoting their proliferation [89]. Another aspect to consider is that culture media are often deficient in antioxidants especially in tocopherols and ascorbate. The fact that the cells are deprived of these antioxidants in the culture medium can lead to a false interpretation of the beneficial effects of the added antioxidant. In other words, antioxidants may appear beneficial when added to cells in culture, but it is because the deficiency is being corrected and not because there is a real "super antioxidant" effect.

CONCLUSION

Analysis of the mechanisms by which flavonoids exert their antioxidant activity as trapping agents of reactive oxygen species (ROS) has been performed. However, after the metabolic process and due to the high enzymatic reactivity of reactive oxygen species, flavonoids become prooxidant agents or promoters of the production of reactive oxygen species, mechanisms that have also been analyzed.

REFERENCES

[1]   Taiz, L. *Zeiger, E. Plant physiology,* 2nd ed; Sinauer Associates, Inc.: USA, **1998**.

[2]   Ross, J.A.; Kasum, C.M. Dietary flavonoids: bioavailability, metabolic effects, and safety. *Annu. Rev. Nutr.,* **2002**, *22*, 19-34.
        [http://dx.doi.org/10.1146/annurev.nutr.22.111401.144957] [PMID: 12055336]

[3]   Veitch, N.C.; Grayer, R.J. Flavonoids and their glycosides, including anthocyanins. *Nat. Prod. Rep.,* **2008**, *25*(3), 555-611.
        [http://dx.doi.org/10.1039/b718040n] [PMID: 18497898]

[4]   Williams, C.A.; Grayer, R.J. Anthocyanins and other flavonoids. *Nat. Prod. Rep.,* **2004**, *21*(4), 539-573.
        [http://dx.doi.org/10.1039/b311404j] [PMID: 15282635]

[5]   S.; González-Gallego, J.; Culebras, J. M.; Muñón, M. J. Los flavonoides: propiedades y acciones antioxidantes. *Nutr. Hosp.,* **2002**, *XVII*, 6-, 271-278.

[6]   Arora, A.; Nair, M.G.; Strasburg, G.M. Structure-activity relationships for antioxidant activities of a series of flavonoids in a liposomal system. *Free Radic. Biol. Med.,* **1998**, *24*(9), 1355-1363.
        [http://dx.doi.org/10.1016/S0891-5849(97)00458-9] [PMID: 9641252]

[7]   Chaves, N.; Escudero, J.C.; Gutiérrez-Merino, C. Seasonal variation of exudate of *Cistus ladanifer. J. Chem. Ecol.,* **1993**, *19*(11), 2577-2591.
        [http://dx.doi.org/10.1007/BF00980692] [PMID: 24248712]

[8]   Maurya, R.; Yadav, P.P. Furanoflavonoids: an overview. *Nat. Prod. Rep.,* **2005**, *22*(3), 400-424.

[http://dx.doi.org/10.1039/b505071p] [PMID: 16010348]

[9]   Yao, L.H.; Jiang, Y.M.; Shi, J.; Tomás-Barberán, F.A.; Datta, N.; Singanusong, R.; Chen, S.S. Flavonoids in food and their health benefits. *Plant Foods Hum. Nutr.,* **2004**, *59*(3), 113-122. [http://dx.doi.org/10.1007/s11130-004-0049-7] [PMID: 15678717]

[10]   Weinreb, O.; Mandel, S.; Amit, T.; Youdim, M.B. Neurological mechanisms of green tea polyphenols in Alzheimer's and Parkinson's diseases. *J. Nutr. Biochem.,* **2004**, *15*(9), 506-516. [http://dx.doi.org/10.1016/j.jnutbio.2004.05.002] [PMID: 15350981]

[11]   Harborne, J.B.; Williams, C.A. Anthocyanins and other flavonoids. *Nat. Prod. Rep.,* **2001**, *18*(3), 310-333. [http://dx.doi.org/10.1039/b006257j] [PMID: 11476484]

[12]   Veitch, N.C. Isoflavonoids of the leguminosae. *Nat. Prod. Rep.,* **2007**, *24*(2), 417-464. [http://dx.doi.org/10.1039/b511238a] [PMID: 17390003]

[13]   Erlund, I. Review of the flavonoids quercetin, hesperetin and naringenin. Dietary sources, bioactivities, bioavailability, and epidemiology. *Nutr. Res.,* **2004**, *24*, 851-874. [http://dx.doi.org/10.1016/j.nutres.2004.07.005]

[14]   Reynaud, J.; Guilet, D.; Terreux, R.; Lussignol, M.; Walchshofer, N. Isoflavonoids in non-leguminous families: an update. *Nat. Prod. Rep.,* **2005**, *22*(4), 504-515. [http://dx.doi.org/10.1039/b416248j] [PMID: 16047048]

[15]   Sosa, T.; Chaves, N.; Alias, J.C.; Escudero, J.C.; Henao, F.; Gutiérrez-Merino, C. Inhibition of mouth skeletal muscle relaxation by flavonoids of *Cistus ladanifer* L.: a plant defense mechanism against herbivores. *J. Chem. Ecol.,* **2004**, *30*(6), 1087-1101. [http://dx.doi.org/10.1023/B:JOEC.0000030265.45127.08] [PMID: 15303316]

[16]   Harborne, J.B.; Williams, C.A. Advances in flavonoid research since 1992. *Phytochemistry,* **2000**, *55*(6), 481-504. [http://dx.doi.org/10.1016/S0031-9422(00)00235-1] [PMID: 11130659]

[17]   Wang, L.; Tu, Y.C.; Lian, T.W.; Hung, J.T.; Yen, J.H.; Wu, M.J. Distinctive antioxidant and antiinflammatory effects of flavonols. *J. Agric. Food Chem.,* **2006**, *54*(26), 9798-9804. [http://dx.doi.org/10.1021/jf0620719] [PMID: 17177504]

[18]   Suomela, J.P.; Ahotupa, M.; Yang, B.; Vasankari, T.; Kallio, H. Absorption of flavonols derived from sea buckthorn (Hippophaë rhamnoides L.) and their effect on emerging risk factors for cardiovascular disease in humans. *J. Agric. Food Chem.,* **2006**, *54*(19), 7364-7369. [http://dx.doi.org/10.1021/jf061889r] [PMID: 16968106]

[19]   González-Gallego, J.; Sánchez-Campos, S.; Tuñón, M.J. Anti-inflammatory properties of dietary flavonoids. *Nutr. Hosp.,* **2007**, *22*(3), 287-293. [PMID: 17612370]

[20]   Edwards, R.L.; Lyon, T.; Litwin, S.E.; Rabovsky, A.; Symons, J.D.; Jalili, T. Quercetin reduces blood pressure in hypertensive subjects. *J. Nutr.,* **2007**, *137*(11), 2405-2411. [http://dx.doi.org/10.1093/jn/137.11.2405] [PMID: 17951477]

[21]   Landis-Piwowar, K.R.; Huo, C.; Chen, D.; Milacic, V.; Shi, G.; Chan, T.H.; Dou, Q.P. A novel prodrug of the green tea polyphenol (-)-epigallocatechin-3-gallate as a potential anticancer agent. *Cancer Res.,* **2007**, *67*(9), 4303-4310. [http://dx.doi.org/10.1158/0008-5472.CAN-06-4699] [PMID: 17483343]

[22]   Cushnie, T.P.T.; Lamb, A.J. Antimicrobial activity of flavonoids. *Int. J. Antimicrob. Agents,* **2005**, *26*(5), 343-356. [http://dx.doi.org/10.1016/j.ijantimicag.2005.09.002] [PMID: 16323269]

[23]   Rao, Y.K.; Fang, S.H.; Tzeng, Y.M. Antiinflammatory activities of flavonoids and a triterpene caffeate isolated from Bauhinia variegata. *Phytother. Res.,* **2008**, *22*(7), 957-962. [http://dx.doi.org/10.1002/ptr.2448] [PMID: 18384188]

[24] Lago, J.H.G.; Toledo-Arruda, A.C.; Mernak, M.; Barrosa, K.H.; Martins, M.A.; Tibério, I.F.L.C.; Prado, C.M. Structure-activity association of flavonoids in lung diseases. *Molecules,* **2014**, *19*(3), 3570-3595.
[http://dx.doi.org/10.3390/molecules19033570] [PMID: 24662074]

[25] Mattioli, V.; Zanolin, M.E.; Cazzoletti, L.; Bono, R.; Cerveri, I.; Ferrari, M.; Pirina, P.; Garcia-Larsen, V. Dietary flavonoids and respiratory diseases: a population-based multi-case-control study in Italian adults. *Public Health Nutr.,* **2020**, *23*(14), 2548-2556.
[http://dx.doi.org/10.1017/S1368980019003562] [PMID: 31996276]

[26] Jeong, J.M.; Choi, C.H.; Kang, S.K.; Lee, I.H.; Lee, J.Y.; Jung, H. Antioxidant and chemosensitizing effects of flavonoids with hydroxy and/or methoxy groups and structure-activity relationship. *J. Pharm. Pharm. Sci.* **2007**, *10*(4), 537-546.
[http://dx.doi.org/10.18433/J3KW2Z] [PMID: 18261373]

[27] Espíndola, C. Some ways for the synthesis of chalcones - new ways for the synthesis of flavon-3- ols. *Mini Rev. Org. Chem.,* **2020**, *17*, 647-673.
[http://dx.doi.org/10.2174/1570193X16666190919111252]

[28] Amić, D.; Davidović-Amić, D.; Běslo, D.; Rastija, V.; Lučić, B.; Trinajstić, N. SAR and QSAR of the antioxidant activity of flavonoids. *Curr. Med. Chem.,* **2007**, *14*(7), 827-845.
[http://dx.doi.org/10.2174/092986707780090954] [PMID: 17346166]

[29] Sichel, G.; Corsaro, C.; Scalia, M.; Di Bilio, A.J.; Bonomo, R.P. *In vitro* scavenger activity of some flavonoids and melanins against O2-(.). *Free Radic. Biol. Med.,* **1991**, *11*(1), 1-8.
[http://dx.doi.org/10.1016/0891-5849(91)90181-2] [PMID: 1657731]

[30] Heim, K.E.; Tagliaferro, A.R.; Bobilya, D.J. Flavonoid antioxidants: chemistry, metabolism and structure-activity relationships. *J. Nutr. Biochem.,* **2002**, *13*(10), 572-584.
[http://dx.doi.org/10.1016/S0955-2863(02)00208-5] [PMID: 12550068]

[31] Potapovich, A.I.; Kostyuk, V.A. Comparative study of antioxidant properties and cytoprotective activity of flavonoids. *Biochemistry (Mosc.),* **2003**, *68*(5), 514-519. [Moscow].
[http://dx.doi.org/10.1023/A:1023947424341] [PMID: 12882632]

[32] Husain, S.R.; Cillard, J.; Cillard, P. Hidroxyl radical scavenging activity of flavonoids. *Phytochemistry,* **1987**, *26*, 2489-2491.
[http://dx.doi.org/10.1016/S0031-9422(00)83860-1]

[33] Kowaltowski, A.J.; de Souza-Pinto, N.C.; Castilho, R.F.; Vercesi, A.E. Mitochondria and reactive oxygen species. *Free Radic. Biol. Med.,* **2009**, *47*(4), 333-343.
[http://dx.doi.org/10.1016/j.freeradbiomed.2009.05.004] [PMID: 19427899]

[34] Saija, A.; Scalese, M.; Lanza, M.; Marzullo, D.; Bonina, F.; Castelli, F. Flavonoids as antioxidant agents: importance of their interaction with biomembranes. *Free Radic. Biol. Med.,* **1995**, *19*(4), 481-486.
[http://dx.doi.org/10.1016/0891-5849(94)00240-K] [PMID: 7590397]

[35] Silva, M.M.; Santos, M.R.; Caroço, G.; Rocha, R.; Justino, G.; Mira, L. Structure-antioxidant activity relationships of flavonoids: a re-examination. *Free Radic. Res.,* **2002**, *36*(11), 1219-1227.
[http://dx.doi.org/10.1080/198-1071576021000016472] [PMID: 12592674]

[36] Santos, M.R.; Mira, L. Protection by flavonoids against the peroxynitrite-mediated oxidation of dihydrorhodamine. *Free Radic. Res.,* **2004**, *38*(9), 1011-1018.
[http://dx.doi.org/10.1080/10715760400003384] [PMID: 15621720]

[37] Sutherland, B.A.; Rahman, R.M.; Appleton, I. Mechanisms of action of green tea catechins, with a focus on ischemia-induced neurodegeneration. *J. Nutr. Biochem.,* **2006**, *17*(5), 291-306.
[http://dx.doi.org/10.1016/j.jnutbio.2005.10.005] [PMID: 16443357]

[38] Lambert, J.D.; Elias, R.J. The antioxidant and pro-oxidant activities of green tea polyphenols: a role in cancer prevention. *Arch. Biochem. Biophys.,* **2010**, *501*(1), 65-72. [Review].

[http://dx.doi.org/10.1016/j.abb.2010.06.013] [PMID: 20558130]

[39] Qian, S.Y.; Buettner, G.R. Iron and dioxygen chemistry is an important route to initiation of biological free radical oxidations: an electron paramagnetic resonance spin trapping study. *Free Radic. Biol. Med.,* **1999**, *26*(11-12), 1447-1456.
[http://dx.doi.org/10.1016/S0891-5849(99)00002-7] [PMID: 10401608]

[40] Trouillas, P.; Marsal, P.; Siri, D.; Lazzaroni, R.; Duroux, J-L. A DFT study of the reactivity of OH groups in quercetin and taxifolin antioxidants: The specificity of the 3-OH site. *Food Chem.,* **2006**, *97*, 679-688.
[http://dx.doi.org/10.1016/j.foodchem.2005.05.042]

[41] Afanas'ev, I.B.; Dorozhko, A.I.; Brodskii, A.V.; Kostyuk, V.A.; Potapovitch, A.I. Chelating and free radical scavenging mechanisms of inhibitory action of rutin and quercetin in lipid peroxidation. *Biochem. Pharmacol.,* **1989**, *38*(11), 1763-1769.
[http://dx.doi.org/10.1016/0006-2952(89)90410-3] [PMID: 2735934]

[42] Dhaouadi, Z.; Nsangou, M.; Garrab, N.; Anouar, E.H.; Marakchi, K.; Lahmar, S. DFT study of the reaction of quercetin with O^{-}_{2} and. OH radicals. *J. Mol. Struct. THEOCHEM,* **2009**, *904*, 35-42.
[http://dx.doi.org/10.1016/j.theochem.2009.02.034]

[43] Ghiasi, M.; Heravi, M.M. Quantum mechanical study of antioxidative ability and antioxidative mechanism of rutin (vitamin P) in solution. *Carbohydr. Res.,* **2011**, *346*(6), 739-744.
[http://dx.doi.org/10.1016/j.carres.2011.01.021] [PMID: 21397896]

[44] Olson, J.S.; Ballow, D.P.; Palmer, G.; Massey, V. The reaction of xanthine oxidase with molecular oxygen. *J. Biol. Chem.,* **1974**, *249*(14), 4350-4362.
[http://dx.doi.org/10.1016/S0021-9258(19)42427-7] [PMID: 4367214]

[45] Olson, J.S.; Ballou, D.P.; Palmer, G.; Massey, V. The mechanism of action of xanthine oxidase. *J. Biol. Chem.,* **1974**, *249*(14), 4363-4382.
[http://dx.doi.org/10.1016/S0021-9258(19)42428-9] [PMID: 4367215]

[46] Day, A.J.; Bao, Y.; Morgan, M.R.A.; Williamson, G. Conjugation position of quercetin glucuronides and effect on biological activity. *Free Radic. Biol. Med.,* **2000**, *29*(12), 1234-1243.
[http://dx.doi.org/10.1016/S0891-5849(00)00416-0] [PMID: 11118813]

[47] Krishnaiah, D.; Sarbatly, R.; Nithyanandam, R. A review of the antioxidant potential of medicinal plant species. *Food Bioprod. Process.,* **2011**, *89*, 217-233.
[http://dx.doi.org/10.1016/j.fbp.2010.04.008]

[48] Cao, G.; Sofic, E.; Prior, R.L. Antioxidant and prooxidant behavior of flavonoids: structure-activity relationships. *Free Radic. Biol. Med.,* **1997**, *22*(5), 749-760.
[http://dx.doi.org/10.1016/S0891-5849(96)00351-6] [PMID: 9119242]

[49] Prior, R.L.; Wu, X.; Schaich, K. Standardized methods for the determination of antioxidant capacity and phenolics in foods and dietary supplements. *J. Agric. Food Chem.,* **2005**, *53*(10), 4290-4302.
[http://dx.doi.org/10.1021/jf0502698] [PMID: 15884874]

[50] Muzolf, M.; Szymusiak, H.; Gliszczyńska-Swigło, A.; Rietjens, I.M.; Tyrakowska, B. pH-Dependent radical scavenging capacity of green tea catechins. *J. Agric. Food Chem.,* **2008**, *56*(3), 816-823.
[http://dx.doi.org/10.1021/jf0712189] [PMID: 18179168]

[51] Aruoma, O.I. Methodological considerations for characterizing potential antioxidant actions of bioactive components in plant foods. *Mutat. Res.,* **2003**, *523-524*, 9-20.
[http://dx.doi.org/10.1016/S0027-5107(02)00317-2] [PMID: 12628499]

[52] Mata-Bilbao, M. de L.; Andrés-Lacueva, C.; Roura, E.; Jáuregui, O.; Torre, C.; Lamuela-Raventós, R.M. A new LC/MS/MS rapid and sensitive method for the determination of green tea catechins and their metabolites in biological samples. *J. Agric. Food Chem.,* **2007**, *55*(22), 8857-8863.
[http://dx.doi.org/10.1021/jf0713962] [PMID: 17902624]

[53] Kotani, A.; Takahashi, K.; Hakamata, H.; Kojima, S.; Kusu, F. Attomole catechins determination by

capillary liquid chromatography with electrochemical detection. *Anal. Sci.,* **2007**, *23*(2), 157-163.
[http://dx.doi.org/10.2116/analsci.23.157] [PMID: 17297226]

[54]   Vidal, G.J.; Veciana, M.J. Espectroscopía de resonancia paramagnética electrónica aplicada al estudio de radicales libres orgánicos. *Cuarta escuela de resonancia paramagnética electrónica*; Universidad de Alicante – España, **2004**.

[55]   Moon, Y.J.; Wang, L.; DiCenzo, R.; Morris, M.E. Quercetin pharmacokinetics in humans. *Biopharm. Drug Dispos.,* **2008**, *29*(4), 205-217.
[http://dx.doi.org/10.1002/bdd.605] [PMID: 18241083]

[56]   Prasain, J.K.; Wang, C-C.; Barnes, S. Mass spectrometric methods for the determination of flavonoids in biological samples. *Free Radic. Biol. Med.,* **2004**, *37*(9), 1324-1350.
[http://dx.doi.org/10.1016/j.freeradbiomed.2004.07.026] [PMID: 15454273]

[57]   Griendling, K.K.; Touyz, R.M.; Zweier, J.L.; Dikalov, S.; Chilian, W.; Chen, Y-R.; Harrison, D.G.; Bhatnagar, A. Measurement of reactive oxygen species, reactive nitrogen species, and redox-dependent signaling in the cardiovascular system: A scientific statement from the american heart Association. *Circ. Res.,* **2016**, *119*(5), e39-e75.
[http://dx.doi.org/10.1161/RES.0000000000000110] [PMID: 27418630]

[58]   Jiang, L.; Yanase, E.; Mori, T.; Kurata, K.; Toyama, M.; Tsuchiya, A.; Yamauchi, K.; Mitsunaga, T.; Iwahashi, H.; Takahashi, J. Relationship between flavonoid structure and reactive oxygen species generation upon ultraviolet and X-ray irradiation. *J. Photochem. Photobiol. Chem.,* **2019**, *384*112044
[http://dx.doi.org/10.1016/j.jphotochem.2019.112044]

[59]   D'Archivio, M.; Filesi, C.; Di Benedetto, R.; Gargiulo, R.; Giovannini, C.; Masella, R. Polyphenols, dietary sources and bioavailability. *Ann. Ist. Super. Sanita,* **2007**, *43*(4), 348-361.
[PMID: 18209268]

[60]   Havsteen, B.H. The biochemistry and medical significance of the flavonoids. *Pharmacol. Ther.,* **2002**, *96*(2-3), 67-202.
[http://dx.doi.org/10.1016/S0163-7258(02)00298-X] [PMID: 12453566]

[61]   Miyake, Y.; Sakurai, C.; Usuda, M.; Fukumoto, S.; Hiramitsu, M.; Sakaida, K.; Osawa, T.; Kondo, K. Difference in plasma metabolite concentration after ingestion of lemon flavonoids and their aglycones in humans. *J. Nutr. Sci. Vitaminol. (Tokyo),* **2006**, *52*(1), 54-60.
[http://dx.doi.org/10.3177/jnsv.52.54] [PMID: 16637230]

[62]   Lee, M.J.; Wang, Z.Y.; Li, H.; Chen, L.; Sun, Y.; Gobbo, S.; Balentine, D.A.; Yang, C.S. Analysis of plasma and urinary tea polyphenols in human subjects. *Cancer Epidemiol. Biomarkers Prev.,* **1995**, *4*(4), 393-399.
[PMID: 7655336]

[63]   Dangles, O.; Dufour, C.; Manach, C.; Morand, C.; Remesy, C. Binding of flavonoids to plasma proteins. *Methods Enzymol.,* **2001**, *335*, 319-333.
[http://dx.doi.org/10.1016/S0076-6879(01)35254-0] [PMID: 11400381]

[64]   Manach, C.; Morand, C.; Gil-Izquierdo, A.; Bouteloup-Demange, C.; Rémésy, C. Bioavailability in humans of the flavanones hesperidin and narirutin after the ingestion of two doses of orange juice. *Eur. J. Clin. Nutr.,* **2003**, *57*(2), 235-242.
[http://dx.doi.org/10.1038/sj.ejcn.1601547] [PMID: 12571654]

[65]   Kanaze, F.I.; Bounartzi, M.I.; Georgarakis, M.; Niopas, I. Pharmacokinetics of the citrus flavanone aglycones hesperetin and naringenin after single oral administration in human subjects. *Eur. J. Clin. Nutr.,* **2007**, *61*(4), 472-477.
[http://dx.doi.org/10.1038/sj.ejcn.1602543] [PMID: 17047689]

[66]   Manach, C.; Donovan, J.L. Pharmacokinetics and metabolism of dietary flavonoids in humans. *Free Radic. Res.,* **2004**, *38*(8), 771-785.
[http://dx.doi.org/10.1080/10715760410001727858] [PMID: 15493450]

[67] Vergne, S.; Titier, K.; Bernard, V.; Asselineau, J.; Durand, M.; Lamothe, V.; Potier, M.; Perez, P.; Demotes-Mainard, J.; Chantre, P.; Moore, N.; Bennetau-Pelissero, C.; Sauvant, P. Bioavailability and urinary excretion of isoflavones in humans: effects of soy-based supplements formulation and equol production. *J. Pharm. Biomed. Anal.,* **2007**, *43*(4), 1488-1494.
[http://dx.doi.org/10.1016/j.jpba.2006.10.006] [PMID: 17110073]

[68] Lotito, S.B.; Frei, B. Consumption of flavonoid-rich foods and increased plasma antioxidant capacity in humans: cause, consequence, or epiphenomenon? *Free Radic. Biol. Med.,* **2006**, *41*(12), 1727-1746.
[http://dx.doi.org/10.1016/j.freeradbiomed.2006.04.033] [PMID: 17157175]

[69] Manach, C.; Scalbert, A.; Morand, C.; Rémésy, C.; Jiménez, L. Polyphenols: food sources and bioavailability. *Am. J. Clin. Nutr.,* **2004**, *79*(5), 727-747.
[http://dx.doi.org/10.1093/ajcn/79.5.727] [PMID: 15113710]

[70] Gardana, C.; Guarnieri, S.; Riso, P.; Simonetti, P.; Porrini, M. Flavanone plasma pharmacokinetics from blood orange juice in human subjects. *Br. J. Nutr.,* **2007**, *98*(1), 165-172.
[http://dx.doi.org/10.1017/S0007114507699358] [PMID: 17367568]

[71] Fraga, C.G. Plant polyphenols: how to translate their *in vitro* antioxidant actions to *in vivo* conditions. *IUBMB Life,* **2007**, *59*(4-5), 308-315.
[http://dx.doi.org/10.1080/15216540701230529] [PMID: 17505970]

[72] Justino, G.C.; Santos, M.R.; Canário, S.; Borges, C.; Florêncio, M.H.; Mira, L. Plasma quercetin metabolites: structure-antioxidant activity relationships. *Arch. Biochem. Biophys.,* **2004**, *432*(1), 109-121.
[http://dx.doi.org/10.1016/j.abb.2004.09.007] [PMID: 15519302]

[73] Pollard, S.E.; Kuhnle, G.G.; Vauzour, D.; Vafeiadou, K.; Tzounis, X.; Whiteman, M.; Rice-Evans, C.; Spencer, J.P.E. The reaction of flavonoid metabolites with peroxynitrite. *Biochem. Biophys. Res. Commun.,* **2006**, *350*(4), 960-968.
[http://dx.doi.org/10.1016/j.bbrc.2006.09.131] [PMID: 17045238]

[74] Ottaviani, J.I.; Momma, T.Y.; Heiss, C.; Kwik-Uribe, C.; Schroeter, H.; Keen, C.L. The stereochemical configuration of flavanols influences the level and metabolism of flavanols in humans and their biological activity *in vivo. Free Radic. Biol. Med.,* **2011**, *50*(2), 237-244.
[http://dx.doi.org/10.1016/j.freeradbiomed.2010.11.005] [PMID: 21074608]

[75] Bjelakovic, G.; Nikolova, D.; Gluud, L.L.; Simonetti, R.G.; Gluud, C. Mortality in randomized trials of antioxidant supplements for primary and secondary prevention: systematic review and meta-analysis. *JAMA,* **2007**, *297*(8), 842-857.
[http://dx.doi.org/10.1001/jama.297.8.842] [PMID: 17327526]

[76] Metodiewa, D.; Jaiswal, A.K.; Cenas, N.; Dickancaité, E.; Segura-Aguilar, J. Quercetin may act as a cytotoxic prooxidant after its metabolic activation to semiquinone and quinoidal product. *Free Radic. Biol. Med.,* **1999**, *26*(1-2), 107-116.
[http://dx.doi.org/10.1016/S0891-5849(98)00167-1] [PMID: 9890646]

[77] Land, E.J.; Swallow, A.J. One-electron reactions in biochemical systems as studied by pulse radiolysis. 3. Ubiquinone. *J. Biol. Chem.,* **1970**, *245*(8), 1890-1894.
[http://dx.doi.org/10.1016/S0021-9258(18)63182-5] [PMID: 5440834]

[78] Bors, W.; Michel, C.; Schikora, S. Interaction of flavonoids with ascorbate and determination of their univalent redox potentials: a pulse radiolysis study. *Free Radic. Biol. Med.,* **1995**, *19*(1), 45-52.
[http://dx.doi.org/10.1016/0891-5849(95)00011-L] [PMID: 7635358]

[79] Jovanovic, S.V.; Steenken, S.; Tosic, M.; Marjanovic, B.; Simic, M.G. Flavonoids as antioxidants. *J. Am. Chem. Soc.,* **1994**, *116*, 4846-4851.
[http://dx.doi.org/10.1021/ja00090a032]

[80] Forni, L.G.; Willson, R.L. Thiyl and phenoxyl free radicals and NADH. Direct observation of one-electron oxidation. *Biochem. J.,* **1986**, *240*(3), 897-903.

[http://dx.doi.org/10.1042/bj2400897] [PMID: 3827879]

[81]   Land, E.J.; Swallow, A.J. One-electron reactions in biochemical systems as studied by pulse radiolysis. IV. Oxidation of dihydronicotinamide-adenine dinucleotide. *Biochim. Biophys. Acta,* **1971**, *234*(1), 34-42.
[http://dx.doi.org/10.1016/0005-2728(71)90126-5] [PMID: 4327080]

[82]   Sudhar, P.S.; Armstrong, D.A. Redox potential of some sulfur containing radicals. *J. Phys. Chem.,* **1990**, *94*, 5915-5917.

[83]   Hodnick, W.F.; Milosavljević, E.B.; Nelson, J.H.; Pardini, R.S. Electrochemistry of flavonoids. Relationships between redox potentials, inhibition of mitochondrial respiration, and production of oxygen radicals by flavonoids. *Biochem. Pharmacol.,* **1988**, *37*(13), 2607-2611.
[http://dx.doi.org/10.1016/0006-2952(88)90253-5] [PMID: 3390220]

[84]   Chan, T.; Galati, G.; O'Brien, P.J. Oxygen activation during peroxidase catalysed metabolism of flavones or flavanones. *Chem. Biol. Interact.,* **1999**, *122*(1), 15-25.
[http://dx.doi.org/10.1016/S0009-2797(99)00103-9] [PMID: 10475612]

[85]   Galati, G.; Sabzevari, O.; Wilson, J.X.; O'Brien, P.J. Prooxidant activity and cellular effects of the phenoxyl radicals of dietary flavonoids and other polyphenolics. *Toxicology,* **2002**, *177*(1), 91-104.
[http://dx.doi.org/10.1016/S0300-483X(02)00198-1] [PMID: 12126798]

[86]   Galati, G.; Moridani, M.Y.; Chan, T.S.; O'Brien, P.J. Peroxidative metabolism of apigenin and naringenin *versus* luteolin and quercetin: glutathione oxidation and conjugation. *Free Radic. Biol. Med.,* **2001**, *30*(4), 370-382.
[http://dx.doi.org/10.1016/S0891-5849(00)00481-0] [PMID: 11182292]

[87]   Severino, J.F.; Goodman, B.A.; Kay, C.W.; Stolze, K.; Tunega, D.; Reichenauer, T.G.; Pirker, K.F. Free radicals generated during oxidation of green tea polyphenols: electron paramagnetic resonance spectroscopy combined with density functional theory calculations. *Free Radic. Biol. Med.,* **2009**, *46*(8), 1076-1088.
[http://dx.doi.org/10.1016/j.freeradbiomed.2009.01.004] [PMID: 19439236]

[88]   Boots, A.W.; Li, H.; Schins, R.P.; Duffin, R.; Heemskerk, J.W.; Bast, A.; Haenen, G.R. The quercetin paradox. *Toxicol. Appl. Pharmacol.,* **2007**, *222*(1), 89-96.
[http://dx.doi.org/10.1016/j.taap.2007.04.004] [PMID: 17537471]

[89]   Halliwell, B. Are polyphenols antioxidants or pro-oxidants? What do we learn from cell culture and *in vivo* studies? *Arch. Biochem. Biophys.,* **2008**, *476*(2), 107-112.
[http://dx.doi.org/10.1016/j.abb.2008.01.028] [PMID: 18284912]

CHAPTER 6

Oxygen Availability

Abstract: The ability to extract oxygen from the environment and deliver it to each cell of the multicellular organism through metabolism in time was the main development of organisms during evolution. The life of living organisms is absolutely dependent on oxygen supplementation for respiration, a process by which cells produce ATP to obtain energy energy from controlled reactions of hydrogen with oxygen to produce water. Depending on cell type, function and biological state, cells have a broad range in oxygen utilization. Here there is evidence on the considerable increase of reactive oxygen species-ROS caused by metabolic alterations, as response the low levels of oxygen available-hypoxia that occur in human cells.

Keywords: Glucose Regulating Proteins (GRPs), Heat Shock Proteins (HSPs), Hipoxic Conditions, Hypoxia-Associated Proteins (HAPS) and Oxygen Regulating Proteins (ORPs), Oxygen Consumption Rates (OCRs), Oxygen Levels in the Blood, Respiratory Insufficiency.

1. REACTIVE OXYGEN SPECIES-ROS IN HYPOXIA

The human body utilizes oxygen to extract approximately 2,550 calories from food to provide for daily energy needs. For example, 10.4 MJ are required for a 20-year-old person and 70 kg, this consumption requires approximately 22 moles of O_2 per day, or 2.5×10^{-4} mol s^{-1}. Thus, the oxygen consumption rate for the 70 kg subject is 3.6×10^{-9} mol s^{-1} g^{-1}. Furthermore, if this 70 kg human contains 1×10^{14} cells, the oxygen utilization rate per cell would be 2.5×10^{-18} mol cell^{-1}s^{-1}.

Oxygen utilization rate by cells depends on cell type, function, and biological state. Thus, the oxygen utilization rate by a red blood cell without mitochondria, which depends entirely on glycolysis to supply its energy needs instead of respiration, is very different from the oxygen utilization rate by a hepatocyte containing a number of mitochondria of the order 10^3.

The majority of the O_2 utilized in mitochondrial respiration presents a four-electron reduction to produce water (reaction 2). However, $O^{\cdot-}_2$ superoxide is also formed by the one-electron reduction of a small fraction of this oxygen, corresponding to $\approx$ 1% or less of the Oxygen Consumption Rate-OCR. Therefore, in the electron transport of mitochondria *In vivo*, this O_2 reduction could be less than this.

O^{-}_2 superoxide and H_2O_2 hydrogen peroxide can initiate or contribute to the generation of different pathologies. However, they are species that contribute to the organism's redox biology by establishing a suitable reducing environment in cells and tissues. Oxygen consumption rate-OCR studies are indispensable for understanding the mechanisms by which reactive oxygen species affect redox biology of cells and tissues. For partially reduced species such as superoxide and hydrogen peroxide, this rate corresponds to the absolute upper limit of their potential flux.

Wagner *et al.*, 2011, found that the oxygen consumption rate-OCR depends on the size of the cell and the number of proteins it contains. In addition, to maintain an adequate redox environment in normal and pathological conditions with different oxygen requirements, the cell develops different strategies. In addition, the surface area-volume relationship varies from cell to cell as the volume of the cell also varies. Therefore, different responses are expected when comparing the effect of exposure of a very small bacterium and a large mammalian cell to hydrogen peroxide.

In vitro cells may have different oxygen consumption rates (OCRs) depending on the state of growth and metabolic demand, such as the oxygen consumption is different between cells in growth phase and quiescent cells or differentiated cells. Cells in logarithmic phase may consume oxygen at higher rates than when they are in exponential phase, because when the redox state of extracellular thiols is adjusted in the logarithmic phase, through the pentose cycle a considerable flow of oxygen and a high ATP requirement is indispensable.

In reaction 3 chapter 1, where the NADPH-oxide electrons transfer from a two-electron reducer, NADPH, to oxygen to produce superoxide, they transfer electrons through the membrane *e.g.* when neutrophils are activated, OCR is increased by superoxide produced by Nox. In some cases, most of the oxygen consumed is associated with superoxide production, as occurs in phagocytic cells with a Nox. However, less than 1% of oxygen utilization produces O^{-}_2 superoxide and H_2O_2 hydrogen peroxide, in ATP-producing metabolic processes. Thus, when the OCR is 20 zmol $cell^{-1}s^{-1}$, the rate of O^{-}_2 production would be 200 zmol $cell^{-1}s^{-1}$. In addition, the rate of H_2O_2 production would be 100 zmol $cell^{-1}s^{-1}$ when O^{-}_2 removal is performed by a dismutation catalyzed by SOD.

In most human cells in the resting state, 5% of oxygen pressure is reached.

However, in the mitochondria, oxygen is consumed by *cytochrome c* oxidase, resulting in an oxygen concentration lower than normal, unlike the concentration in the extracellular environment, thus generating an oxygen gradient [2]. Under these low mitochondrial oxygen conditions, most tissues are in hypoxic conditions [3]. This leads to normal development or to pathophysiological conditions in which a reduced oxygen supply caused by respiratory failure or vascular defects can lead to diabetes, inflammatory diseases, cardiovascular or cerebral ischemic disorders and solid tumors.

The living cells, possess the ability to detect low-oxygen levels in the environment and to activate intracellular and extracellular response mechanisms to regulate normal oxygen levels or adapt to new hypoxic conditions is essential. When an organism detects a decrease in oxygen, one of the first mechanisms it develops is the hyperventilation reflex, caused by glomus type I cells activity in the carotid body, which perceive the oxygen level reduction (15%) and activate the response mechanism to hypoxia. However, these cells are insensitive to oxygen levels in the blood.

In the immediate response to low oxygen levels, there is inhibition of K^+ channels, depolarization of the membrane, calcium influx and from synaptic vesicles the release of neurotransmitters. Carotid body sensitivity to hypoxia is modulated by Dopamine, by stimulating the D_2 receptors. Oxygen levels are related with Dopamine metabolism.

When hypoxic conditions are prolonged, compensatory mechanisms such as polycythemia are activated, *i.e.* an increase in the number red blood cells which results in increased blood oxygen levels, that is carried to all tissues. This mechanism is carried out by the glycoprotein EPO-erythropoietin whose expression in conditions of normal oxygen levels is minimal, but in conditions of hypoxia its production is increased mainly in fibroblasts, in interstitial type I cells from the kidney cortex and outer medulla [4].

Glycolysis rate increases while gluconeogenesis and oxidative phosphorylation levels decrease, during the first hours of hypoxia [5]. To stimulate glycolysis, the activity of the insulin-independent glucose transporter type I (GLUT-1) increases. At the same time, ATP deficiency is compensated by increased gene expression and enzymatic activity of glycolytic enzymes.

The expression of specific-stress proteins is another mechanism of adaptation to low oxygen levels in cells: Glucose regulatory proteins (GRP), Hypoxia-associated proteins (HAPS), Oxygen regulatory proteins (ORP) and Heat shock proteins (HSP).

Oxygen molecular O_2 binds to Fe^{+2} of the *Heme* protein causing porphyrin ring translocation of the ferrous atom to occur, such that the ferrous atom is displaced out of the ring by dissociation of the O_2 atom. However, the conformational change that occurs in *heme* molecule from an *oxy* to a *deoxy* affect *heme* protein function. It makes the *heme* protein one of the first sensors of O_2 levels [6].

Cytochrome b protein corresponds to a membrane complex like NAD(P)H phagocytic oxidase that in an O_2-dependent manner generates H_2O_2. The H_2O_2 is an intracellular signalling molecule that regulates the responses of the cell to environmental O_2 levels, so that during hypoxia the reduction of *cytochrome b* decreases the intracellular concentration of H_2O_2 by decreasing the activity of oxidase [7].

During hypoxia due to low intracellular H_2O_2 concentrations, the protein phosphorylation, hydroxyl $\cdot OH$ radical concentration and reduced glutathione (GSH) concentrations decrease. This induces the thiol proteins into their reduced forms, so this change affects the nucleic acid binding of regulatory proteins, influencing gene expression.

Evidence also exists that hypoxia increases mitochondrial ROS causing hypoxic pulmonary vasoconstriction (HPV) and inducing an increase in intracellular pulmonary artery smooth muscle cell (PASMC) Ca^{2+} [8, 9] as well as activating NADPH oxidase (NOX) through PKCε [10]. Considering the above, Ward, 2008, proposes that synergism exists between mitochondria and NOx for ROS production involving a positive feedback relationship.

Waypa *et al.*, 2013, found that during acute hypoxia, oxidation decreases in the matrix but increases in the cytosol and intermembrane space. Similarly, they found that depending on the cell sites, there are marked differences in basal redox state. For instance, hypoxia conditions increase the oxidation of thiol groups both in the intermembrane space and in the cytosol and that there are differences in the response of subcellular compartments to hypoxia conditions, as some of them present oxidation, while others do not [12].

Cells also adapt to reduced oxygen levels by inducing HIF (Hypoxia Inducible Factor), which mainly affects cell energy homeostasis by inactivating metabolism, activating anaerobic glycolysis, and inhibiting mitochondrial aerobic metabolism: the TCA cycle and phosphorylation oxidative.

A decrease in the concentration of oxygen considerably reduces the availability of cellular energy, mainly through various mechanisms: a decrease in the concentration of oxygen due to the non-saturation of the substrate by the allosteric modulation of cytochrome *c* oxidase (COX) leading to a decrease in respiration rate. Therefore, the potential of oxidative phosphorylation decreases; however, lactate is produced at a rate of 0.5 mol/mol ATP due to an allosteric increase in phosphofructokinase activity by increased glycolysis, which leads to a decrease in the cellular pH causing the consequent vascular problems and ischemic processes.

It is considered that 1-2% of electron flow through the mitochondrial electron chain produces ROS under normal oxygen supply conditions [13, 4]. However, under conditions of hypoxia ROS levels increase.

Under conditions of normal oxygen supply, in mitochondria electrons released from reduced cofactors (NADH and $FADH_2$) flow through the redox centers of the electron chain to molecular oxygen where a flow of protons is coupled from the matrix to the intermembrane space. The protons return through the F_0 sector of the ATP synthase enzyme complex to the matrix, thus promoting ATP synthesis (see Chapter 2). Next, by adenine nucleotide translocator ATP is transported into the cell cytosol.

Under conditions of moderate hypoxia, electrons escape from the redox centers of the respiratory chain and before arriving at *cytochrome c* oxidase reduce molecular oxygen to the $O^{.-}_2$ superoxide anion radical. Then ATP produced by cytosolic glycolysis enters the mitochondria where it is hydrolyzed by F_1F_0-ATPase with protons released from the mitochondrial matrix, to maintain an adequate $\Delta\Psi_m$.

In the last step of oxidative phosphorylation, ADP phosphorylation is catalyzed by the enzyme F_1F_0 ATPase, the ATP synthase. Under normal oxygen supply conditions, this enzyme synthesizes ATP, but under hypoxia conditions the enzyme F_1F_0 ATPase hydrolyzes ATP and utilizes the energy released to pump protons from the mitochondrial matrix into the intermembrane space coincident with the adenine nucleotide translocator. As such, the mitochondrial membrane potential ($\Delta\Psi_m$) decreases (140 mV, negative inside the matrix), well below its endogenous level. This means that during hypoxia the cytosolic ATP is exchanged for the ADP in the matrix, thus maintaining the physiological mitochondrial potential ($\Delta\Psi_m$).

During hypoxia conditions, to avoid ATP dissipation, the F_1F_0 ATPase enzyme must be tightly controlled. This regulation is performed by the IF1 protein, which is a natural $H^+/\Delta\Psi_m$-dependent protein and binds to the catalytic sector of F1 at low pH values and $\Delta\Psi_m$. When IF1 binds to ATP synthase there is a rapid and reversible inhibition of the F_1F_0 ATPase enzyme which can reach up to 50% of maximal activity [15].

Low concentrations of oxygen in cells or hypoxia cause a decrease in the rate of electron transport, a reduction of $\Delta\Psi_m$ and an increase in ROS and NO synthase production. These factors contribute to the stabilisation of HIF which induces cells' metabolic adaptation to hypoxia and activates the mitophagic process.

Oxygen is the final electron acceptor in Complex IV, *cytochrome c* oxidase, as oxygen has a high affinity for it. It was demonstrated that at pH 7.0, oxidative phosphorylation does not depend on oxygen concentration until values below 20 µM; however, it becomes notoriously dependent when the pH is alkaline, for example, the oxygen concentration at a pH of 7.4 for an average respiratory rate is approximately 0.7 µM [16].

The rate of oxygen consumption remains constant below oxygen concentrations of 15 µM [17]. Thus, conditions of hypoxia can be considered at a range of 5 - 0.5% oxygen which corresponds to a concentration range of 46 - 4.6 µM oxygen in cell culture.

Numerous studies (Lakey *et al.*, 2016) [18], have demonstrated that most respiratory problems are caused by an excess of ROS production due to high levels of environmental contaminants. The reaction of atmospheric ozone and reactive oxygen species - ROS with antioxidants such as reduced glutathione, ascorbate, and α-tocopherol and with surfactants produce organic oxidants. Also, redox components such as iron and copper ions and quinones generate and sustain oxidative reaction cycles that yield ROS and oxidative stress.

CONCLUSION

In the previous chapters of this book, the high enzymatic reactivity of oxygen free radicals - ROS under normal oxygen supply has been analyzed. This chapter analyses the behavior of oxygen free radicals under hypoxic conditions, the enzymes and cellular mechanisms involved.

REFERENCES

[1]   Wagner, B.A.; Venkataraman, S.; Buettner, G.R. The rate of oxygen utilization by cells. *Free Radic. Biol. Med.,* **2011**, *51*(3), 700-712.
[http://dx.doi.org/10.1016/j.freeradbiomed.2011.05.024] [PMID: 21664270]

[2]   Solaini, G.; Baracca, A.; Lenaz, G.; Sgarbi, G. Hypoxia and mitochondrial oxidative metabolism. *Biochim. Biophys. Acta,* **2010**, *1797*(6-7), 1171-1177.
[http://dx.doi.org/10.1016/j.bbabio.2010.02.011] [PMID: 20153717]

[3]   Brahimi-Horn, M.C.; Pouysségur, J. Oxygen, a source of life and stress. *FEBS Lett.,* **2007**, *581*(19), 3582-3591.
[http://dx.doi.org/10.1016/j.febslet.2007.06.018] [PMID: 17586500]

[4]   Czyzyk-Krzeska, M.F. Molecular aspects of oxygen sensing in physiological adaptation to hypoxia. *Respir. Physiol.,* **1997**, *110*(2-3), 99-111.

[http://dx.doi.org/10.1016/S0034-5687(97)00076-5] [PMID: 9407604]

[5]   Wölfle, D.; Jungermann, K. Long-term effects of physiological oxygen concentrations on glycolysis and gluconeogenesis in hepatocyte cultures. *Eur. J. Biochem.,* **1985**, *151*(2), 299-303.
[http://dx.doi.org/10.1111/j.1432-1033.1985.tb09100.x] [PMID: 4029136]

[6]   Goldberg, M.A.; Dunning, S.P.; Bunn, H.F. Regulation of the erythropoietin gene: evidence that the oxygen sensor is a heme protein. *Science,* **1988**, *242*(4884), 1412-1415.
[http://dx.doi.org/10.1126/science.2849206] [PMID: 2849206]

[7]   Acker, H. Mechanisms and meaning of cellular oxygen sensing in the organism. *Respir. Physiol.,* **1994**, *95*(1), 1-10.
[http://dx.doi.org/10.1016/0034-5687(94)90043-4] [PMID: 8153448]

[8]   Ward, J.P.T. A twist in the tail: synergism between mitochondria and NADPH oxidase in the hypoxia-induced elevation of reactive oxygen species in pulmonary artery. *Free Radic. Biol. Med.,* **2008**, *45*(9), 1220-1222.
[http://dx.doi.org/10.1016/j.freeradbiomed.2008.08.015] [PMID: 18786634]

[9]   Waypa, G.B.; Guzy, R.; Mungai, P.T.; Mack, M.M.; Marks, J.D.; Roe, M.W.; Schumacker, P.T. Increases in mitochondrial reactive oxygen species trigger hypoxia-induced calcium responses in pulmonary artery smooth muscle cells. *Circ. Res.,* **2006**, *99*(9), 970-978.
[http://dx.doi.org/10.1161/01.RES.0000247068.75808.3f] [PMID: 17008601]

[10]   Rathore, R.; Zheng, Y-M.; Niu, C-F.; Liu, Q-H.; Korde, A.; Ho, Y-S.; Wang, Y.X. Hypoxia activates NADPH oxidase to increase [ROS]i and [Ca2+]i through the mitochondrial ROS-PKCepsilon signaling axis in pulmonary artery smooth muscle cells. *Free Radic. Biol. Med.,* **2008**, *45*(9), 1223-1231.
[http://dx.doi.org/10.1016/j.freeradbiomed.2008.06.012] [PMID: 18638544]

[11]   Waypa, G.B.; Marks, J.D.; Guzy, R.D.; Mungai, P.T.; Schriewer, J.M.; Dokic, D.; Ball, M.K.; Schumacker, P.T. Superoxide generated at mitochondrial complex III triggers acute responses to hypoxia in the pulmonary circulation. *Am. J. Respir. Crit. Care Med.,* **2013**, *187*(4), 424-432.
[http://dx.doi.org/10.1164/rccm.201207-1294OC] [PMID: 23328522]

[12]   Smith, K.A.; Waypa, G.B.; Schumacker, P.T. Redox signaling during hypoxia in mammalian cells. *Redox Biol.,* **2017**, *13*, 228-234.
[http://dx.doi.org/10.1016/j.redox.2017.05.020] [PMID: 28595160]

[13]   Gnaiger, E.; Kuznetsov, A.V. Mitochondrial respiration at low levels of oxygen and *cytochrome c. Biochem. Soc. Trans.,* **2002**, *30*(2), 252-258.
[http://dx.doi.org/10.1042/bst0300252] [PMID: 12023860]

[14]   Turrens, J.F. Superoxide production by the mitochondrial respiratory chain. *Biosci. Rep.,* **1997**, *17*(1), 3-8.
[http://dx.doi.org/10.1023/A:1027374931887] [PMID: 9171915]

[15]   Campanella, M.; Parker, N.; Tan, C.H.; Hall, A.M.; Duchen, M.R. IF(1): setting the pace of the F(1)F(o)-ATP synthase. *Trends Biochem. Sci.,* **2009**, *34*(7), 343-350.
[http://dx.doi.org/10.1016/j.tibs.2009.03.006] [PMID: 19559621]

[16]   Wilson, D.F.; Rumsey, W.L.; Green, T.J.; Vanderkooi, J.M. The oxygen dependence of mitochondrial oxidative phosphorylation measured by a new optical method for measuring oxygen concentration. *J. Biol. Chem.,* **1988**, *263*(6), 2712-2718.
[http://dx.doi.org/10.1016/S0021-9258(18)69126-4] [PMID: 2830260]

[17]   Palacios-Callender, M.; Quintero, M.; Hollis, V.S.; Springett, R.J.; Moncada, S. Endogenous NO regulates superoxide production at low oxygen concentrations by modifying the redox state of cytochrome c oxidase. *Proc. Natl. Acad. Sci. USA,* **2004**, *101*(20), 7630-7635.
[http://dx.doi.org/10.1073/pnas.0401723101] [PMID: 15136725]

[18]   Lakey, P.S.J.; Berkemeier, T.; Tong, H.; Arangio, A.M.; Lucas, K.; Pöschl, U.; Shiraiwa, M. Chemical exposure-response relationship between air pollutants and reactive oxygen species in the human respiratory tract. *Sci. Rep.,* **2016**, *6*, 32916.
[http://dx.doi.org/10.1038/srep32916] [PMID: 27605301]

SUBJECT INDEX

A

Acid(s) 4, 17, 18, 21, 22, 27, 39, 40, 65, 72, 85, 111, 112, 117, 128, 129, 131, 132, 139, 145, 146, 147, 148, 153, 157, 169, 173, 177, 179, 180, 185, 187, 189, 196, 203
   aliphatic 146
   ascorbic 18, 85, 128, 129, 132, 139, 177, 185
   caffeic 189, 203
   cinnamic 189
   fatty 17, 21, 72, 111, 117, 131, 153
   gallic 147, 179
   glucuronic 22, 187
   hyaluronic 22
   hydrolysis 180
   hypochlorous 65, 157
   phenolic 145, 169
   retinoic 112
   shikimic 146, 148
   uric 27, 39, 173
Action, nucleophilic 104
Active enzyme 60, 62
Activity 36, 38, 46, 67, 86, 88, 99, 105, 111, 112, 130, 131, 136, 138, 139, 147, 148, 149, 152, 153, 154, 162, 187, 188, 190, 204, 214, 216
   antiperoxidative 153, 154
   enzymatic 214
   monooxygenase 46
   phosphofructokinase 216
Acylationdesacilation reactions 131
Adenine nucleotide translocase 111, 112
Alkaline autooxidation reactions 204
Aminoxidases 67, 136
Anthocyanins 147, 148, 162, 187, 190
Antioxidant 87, 95, 110, 119, 121, 125, 132, 133, 134, 140, 144, 148, 149, 153, 154, 155, 160, 161, 162, 163, 164, 165, 167, 169, 173, 176, 177, 191, 193

action 87, 95, 110, 119, 121, 125, 134, 140, 144, 153, 160, 161, 162, 163, 164, 165, 167, 169, 173, 176, 177, 191, 193
   activity 132, 133, 148, 149, 154, 155, 161, 162, 163, 164, 165, 169, 176, 177, 191
   and prooxidant reactions 121, 144
   enzymes 87, 95, 110, 125, 140, 173
   plant defense systems 134
Antioxidant defense 95, 125
   enzyme systems 95
   mechanisms 125
Ascorbate 41, 135, 138, 139
   oxidase enzyme 41
   peroxidases 135, 138, 139
ATP 25, 68, 69, 70, 212, 213, 216
   adenine nucleotide translocator 216
   cytosolic 216
   producing metabolic processes 213
   synthase 216
Autocatalytic prooxidant effects 133
Autoxidation 18, 123, 124, 199

B

Biochemical reactions 43
Bioconjugation processes 187
Biogenesis reaction 63
Biological 4, 5, 25, 27, 29, 31, 33, 35, 97
   oxidation 25, 27, 29, 31, 33, 35
   processes 97
   reactions 4, 5
Blood 30, 36, 180, 185, 204, 212, 214
   oxygenated 36
   plasma 204
Body tissues oxygenation 83
Bond dissociation energy (BDEs) 144, 165, 166, 167, 169, 172

C

Calcium channel dysfunction 39

Catalase 18, 30, 31, 33, 34, 86, 87, 94, 95,
    105, 106, 107, 108, 109, 110, 138
  peroxisomal 18
  antioxidant enzyme activity 107
Catalyze 39, 46, 67, 72, 75, 78, 109, 112, 126,
    139, 187
  enzymes UDP-glucuronyltransferases 187
  proteases 39
Catalyzing transacylation reactions 131
Catechol-O-methyl transferase (COMT) 187,
    192, 187
    enzyme 187
Cerebral ischemic disorders 214
Chemical reactions 89
Chiral chromatographic analysis 192
Chromophore 179
Chronic toxicities 86
Coenzyme 17, 27, 28, 110, 113, 132
  electron transporting 27
  fatty acid acyl 17
Combustion theory 4
Crosslinking reaction 62
Cyclooxigenases 149
Cyclooxygenase 38, 155, 173
Cytochrome 43, 44, 49, 75, 76, 77, 99, 215
  oxidase formation 99
  reduction 43, 44, 49, 75, 76, 77, 215
Cytokinesis 83

D

Damage 17, 89, 128, 129, 151
  biomolecules 151
  lipid 129
  mitochondrial 89
  mitochondrial membrane 17
  tissue 128
Death 17, 18, 190, 195
  necrotic cell 17
  reducing cell 195
Decolorization test 178
Decomposition, metal-catalyzed 105
Deglycosylation 186
Degradation processes 183
Dehydrogenase 27, 39, 49
  activity 39
  enzymes 27, 49
Dehydrogenation 29
Delocalization pathway 58
Density functional theory (DFT) 165

Deoxygenases 31
Diabetes mellitus 185
Diacylglycerol lipase 131
Diseases 18, 148, 185, 214
  chronic respiratory 148
  coronary 148
  inflammatory 214
  ischemic heart 185
  lung 148

E

Electron 40, 43, 49, 50, 70, 83, 84, 103, 150,
    166, 181, 182, 184, 204
  paramagnetic resonance (EPR) 49, 70, 103,
    181, 184, 204
  spin resonance (ESR) 40, 50, 83, 84, 150,
    166, 181, 182
  spin resonance spectroscopy 83
  transfer reactions 43
Energy 1, 3, 4, 7, 8, 25, 35, 36, 53, 59, 68, 69,
    165, 167, 212, 215, 216
  cellular 215
  demanding biological redox reaction 3
  free 36, 69
  releasing 4
  transfers 7
  transhydrogenase-bound 35
Energy metabolism 33, 38, 69
  mitochondrial 33
Environment 60, 85, 113, 137, 182, 212, 213,
    214
  chemical 182
  hydrocarbon 113
  oxidative 137
  reducing 213
Enzymatic 31, 38, 50, 90, 94
  kinetic 38
  oxidation processes 90
  reaction mechanisms 94
  reactions 31, 50
Enzyme 30, 55, 96, 108, 109, 137, 139
  galactose oxidase 55
  glutathione peroxidase 30, 109
  oxidoreductase 137
  proteinase 96
  glutathione reductase 108, 139
  superoxide dismutase 108
EPR technique 184
ERS spectroscopy 14

Escherichia coli 96, 113
ESR 41, 76, 84, 151
    spectroscopy 41, 151
    technique 76, 84

F

FAD-containing auxiliary enzyme 72
Fatty carbon radicals 111
Fenton 1, 14, 15, 16, 18, 35, 57, 78, 79, 80,
    81, 82, 85, 111, 164, 178
    reaction 1, 14, 15, 16, 18, 35, 57, 80, 81,
        82, 111
    reagents 14
    type reactions 78, 79, 85
    type reagents 79
    reductive antioxidant power (FRAP) 164,
        178
Ferricyanidin transport 73
Flavanol stereoisomers 192
Flavin 27, 29, 39, 48, 49, 54, 72, 73, 74, 75
    adenine dinucleotide (FAD) 27, 39, 48, 49,
        54, 72, 73
    dehydrogenase 29
    enzymes 73, 74
    microsomal enzymes 75
    reduced enzymes 73
Flavonoids 153, 165, 193
    antioxidant properties 153
    prooxidant activity mechanism 193
    structure-antioxidant activity relationship
        165

G

Galactose oxidase 38, 55, 56, 57, 58, 62
Gas chromatography 181, 182
    mass spectrometry 182
Glucose 108, 187, 212, 214
    6-phosphate dehydrogenase 108
    regulating proteins (GRPs) 212, 214
Glucuronidation reactions 187
Glutathione 126, 128, 138, 139, 155
    peroxidase action 128
    reductase activity 155
    synthase 126
    transferase enzymes 139
    transferases 128, 138
Glycerophosphocholine liposomes 45

Glycolysis 27, 28, 212, 215, 216
    activating anaerobic 215
    cytosolic 216
Glycoprotein 82, 126
Glycosaminoglycan 22
Glycosylation process 187
GM-MS mass spectrometry 181
GSSG reductases catalyze 127

H

Haber-Weiss reaction 8, 35, 79, 125, 137, 138
Heat 135, 212, 214
    shock proteins (HSPs) 212, 214
    stress 135
Heme oxygenase 82
High 172, 181
    occupancy molecular orbital (HOMO) 172
    performance HPLC liquid chromatography
        181
Homeostasis 19
Homogentisate deoxygenase 31
Hydrated electron reactions 11
Hydrolysis 20, 180, 181, 183, 186
    enzymatic 181, 183
Hydroperoxides 20, 40, 45, 86, 130, 139
    lipidic 130
Hypoxia-associated proteins (HAPS) 212, 214
Hypoxic pulmonary vasoconstriction (HPV)
    215

I

Immunoassays 181
Inhibition 111, 149
    redox enzyme's 149
    succinate dehydrogenase 111
Iron 14, 81
    chelator 14
    deficiency 81
    redox reactions 81

L

Lipid 153, 155, 178, 195
    bilayer 153, 195
    hydroperoxidase 155
    hydroperoxide production 178

Lipid peroxidation 20, 21, 22, 35, 81, 85, 89,
        90, 117, 118, 119, 128, 132, 133, 139,
        153, 155, 175
    iron-catalyzed 81
    iron-mediated 35
    non-enzymatic 175
Lipooxygenase inhibition 173
Lipoproteins 132
Liposome disintegration 45
Lipoxygenases 38, 136, 155, 173, 175, 176
    soybean 176
Lipoyl dehydrogenase 85
Liquid chromatography 181
Low-density lipoproteins (LDL) 20, 165, 178

M

Macroxyproteinase 130
Mass spectrometry 182
Meheler reaction 135
Meisenheimer reaction pathway 202
Membrane 83, 109, 115, 130, 131, 195
    dependent processes 195
    lipid peroxidation 83
    phospholipids 115, 130, 131
    protection 109
Mercury reaction 86
Metabolic 22, 144, 186, 187, 205, 212
    alterations 212
    processes 22, 144, 186, 187, 205
Metabolism 3, 19, 23, 25, 28, 30, 34, 36, 67,
        71, 90, 109, 130, 131, 167, 187, 190,
        192, 202
    anthocyanin 187
    fatty acid 67
    mitochondrial 34
    non-enzymatic 90
    oxidative 19, 28, 202
    xenobiotic 71
Metabolite oxidation 26
Microsomal 74, 173
    flavin enzyme 74
    mono oxygenase 173
Molybdenum 48
    enzymes 48
    hydroxylase catalysis 48
Molybdopterin 54
Monodehydroascorbate 41, 43, 74
Monoxygenase activity 72
Myeloperoxidase 40, 65

lysosomal 40

N

NADH 29, 67
    Dehydrogenase 29, 67
    dehydrogenase enzyme 29
NADPH 17, 85, 109
    and lipoyl dehydrogenase 85
    dependent oxidases 17
    quinone oxidoreductase 109
NADPH oxidase 66, 173
    activation in cytochrome 66
    and microsomal mono oxygenase 173
Neurotransmitters 19, 214
Neutrophils NADPH oxidase 66
Nuclear magnetic resonance (NMR) 181, 202

O

Oil, essential 148
Oxidants, organic 217
Oxidation 14, 17, 25, 28, 36, 38, 40, 42, 46,
        55, 60, 69, 81, 86, 101, 135, 158, 163,
        164, 174, 178, 200, 202, 204, 215
    alcohol 55
    electrochemical 163
    electronic 60
    enzymatic 38, 40
    glycolate 135
    metal-catalyzed catechins 158
    peroxidative 42
    reduction 38
    reduction processes 25, 38
    reduction reactions 28, 36
Oxidative 19, 20, 27, 29, 42, 45, 67, 70, 81,
        87, 122, 123, 125, 129, 131, 132, 175
    activity 45
    carboxylation 29
    damage 19, 20, 67, 70, 81, 87, 122, 123,
        125, 129, 131, 132
    enzymes 42
    fragmentation 45
    processes 27, 175
Oxidative stress 20, 67, 83, 84, 87, 94, 97,
        105, 106, 109, 112, 138, 139, 155, 185
    damages macromolecules 20

initiators 155
Oxidized enzyme 56
Oxygen 2, 199, 212, 214,
    atmospheric 2
    activation 199
    regulating proteins (ORPs) 212, 214
Oxygenase enzymes 31, 204

P

Pathways 2, 35, 36, 38, 42, 66, 67, 68, 69, 74,
    132, 134, 135, 144, 146, 149, 190, 192,
    194
    abiotic oxidation 2
    biosynthetic 38
    inner membrane proton conductance 69
    malonic acid 144
    metabolic 67, 68, 134, 192
    photorespiration 135
    shikimic acid 144
    signaling 36
Peroxidase 30, 41, 42, 44, 45, 70, 71, 99, 102,
    155, 195, 202
    action 45
    activity 30, 45, 70, 71, 99, 155
    catalysis 102
    mechanism 195, 202
    oxidase reaction 42
    reaction 41
    system 44
Peroxidation 21, 99, 112, 126, 152, 154, 167,
    175, 178
    enzymatic 175
    ion-induced 154
    iron-stimulated 126
Peroxides 16, 30, 72, 104, 151
    dangerous 30
    deprotoned hydrogen 151
Phagocytosis process 66
Phagolysosome 66, 67
    fusion 67
Phagosome 66, 67
    membrane-enclosed 66
    processing 66
Phenolic oxidation reaction 157
Phospholipid hydroperoxide production 109
Photorespiration process 135
Prooxidant activity 80, 149, 166, 177, 193,
    194, 195, 197, 199

Properties 14, 15, 52, 77, 87, 96, 104, 144,
    148, 149, 151, 152, 153, 155, 165, 171,
    174, 198
    anti-radical 153
    electrochemical 198
    electronic 87, 149
    hydrophilic 165
    hydrophobic 174
    oxidizing 14, 15
    reactive 152
Proteases 130
Protein(s) 17, 18, 19, 20, 22, 23, 55, 56, 66,
    80, 81, 82, 87, 105, 108, 125, 126, 130,
    137, 212, 214, 215
    damaged 130
    heat shock 212, 214
    heme 87, 215
    intracellular iron storage 82
    iron-dependent 80
    stabilization 56
    synthesis 22
    tetrameric 108
Proteolysis 54, 130
    intracellular 130
Proteolytic susceptibility 130
Pulmonary artery smooth muscle cell
    (PASMC) 215

Q

Quantitative structure-activity relationship
    (QSAR) 144, 149
Quercetin176, 191
    inhibitory effects 176
    quinone's reaction 191
Quercetins in chemical systems 195

R

Radical 45, 87, 129, 159
    action 45
    activity, free 159
    formation reactions 129
    production, nickel-mediated free 87
Reaction 9, 19, 125, 126
    free-radical 19, 125, 126
    of oxygen free radicals 9
Reactive oxygen 39, 40, 88, 94, 110, 134, 135,
    137, 194, 212

species production 39, 40
species 88, 94, 110, 134, 135, 137, 194, 212
Redox reactions 3, 38, 45, 81, 126, 127, 130
  catalyse 130
Redox sensitivity 137
Redox system 1, 88, 109
  mitochondrial 88
Reduced glutathione regeneration 85
Reductase 34, 41, 70, 72, 73, 74, 75, 77, 85, 101, 102
Reduction 3, 27, 31, 33, 41, 49, 73, 88, 108, 185
  mechanisms 31
  microbial iron 3
  monoelectronic 33
  oxidized glutathione 108
  processes 27, 49, 88
  reaction 41
  technique 185
Respiration processes 25
Respiratory 214, 217
  failure 214
  problems 217
ROS 38, 110, 136, 139
  in chloroplasts of plant cells 136
  production and membrane 110
  scavenging activity 139
  source enzymes 38

S

Scavengers 139, 173
  radical 173
Scavenging 70, 164
  reaction 164
Signal, cytotoxic 17
Singlet oxygen 7, 40, 106, 107, 132, 138, 164, 186
  producing 40
  oxygen sensor 186
Spectrophotometry 173
Spectroscopy techniques 204
Stoichiometric reaction 166
Stress 34, 97, 135, 138, 139, 140
  abiotic 135, 138, 139, 140
  cold 97
  osmotic 139
Stromal superoxide dismutase enzymes 135

Structure-activity relationship (SAR) 144, 149, 162, 172
Sulfotransferase enzymes catalyze 187
Superoxide 7, 16, 32, 61, 66, 70, 71, 81, 87, 88, 110, 111, 112, 124, 135, 150, 162, 169, 173, 180, 213
  cytoplasmic 87
  mitochondrial 87
Superoxide 7, 20, 32, 40, 45, 70, 94, 95, 98, 99, 122, 138, 158, 172, 173, 180, 199, 201
  dismutase 7, 32, 94, 95, 122, 138, 158, 173, 199, 201
  oxidoreductase 99
  radicals 7, 20, 32, 40, 45, 172, 180
  reductase 98, 99
  scavenging 70
Synechococcus cyanobacteria 97
Synthase 67, 68
Systems 31, 35, 36, 71, 72, 90, 98, 88, 105, 126, 138, 177, 183, 185
  enzymatic 138
  microsomal electron transport 31
  microsomal enzyme 88
  mitochondrial antioxidant 35

T

TEAC 166, 178, 179
  assay 166, 179
  method 178
Tea consumption 183, 190
  green 183
TEAC test 179
Techniques 14, 41, 150, 177, 179, 180, 181, 182, 184, 185, 196, 199
  chromatographic 182
  fluorescence 184
  pulse radiolysis 196, 199
  sensitive spectroscopic 181
  separation 181
  spectrometric 150
Toxicity 18, 32, 33, 78, 81, 85, 89, 126, 134, 193
  mitochondrial 89
  oxygen-free radical 32
Transfer 50, 197
  oxygen-mediated 50
  reaction 197
Transition 158, 78

metal catalytic activity 158
metal toxicity 78
Transport chain reactions 68
Trapped peroxynitrite radicals 164
Tyrosinase mechanism 202
Tyrosine residues 46, 82

U

Ubiquinone 68, 113
    cytochrome 68
    reductase 113
Ultraviolet radiations 8, 155
Urine 182, 193
    analyses 193
    recovery 182
UV radiation 17, 83, 134, 135, 138, 186

V

Vanadium-induced hepatotoxicity 85
Vitamin 94, 118, 119, 122, 125, 128, 129,
        131, 132, 133, 139, 140, 171, 176, 185
    fat-soluble 132
    reactions of 118, 129

W

Water 3, 8, 9, 113, 134
    lipid interface 113
    oxidation 3, 134
    radiolysis 8, 9
Weiss reactions 38, 79

X

Xanthine oxidase (XO) 31, 36, 39, 46, 47, 48,
        49, 50, 52, 54, 77, 173, 174, 176, 180
    catalyze 50
    disulfide enzyme 48
    dehydrogenase 36, 39, 54
    inhibition 173
    lipoxygenase 176
    reaction 77
X-ray crystallography 100, 104

* 9 7 8 9 8 1 5 0 3 6 6 5 7 *